Japanese Aircraft Equipment

Japanese Aircraft Equipment

1940-1945

Robert C. Mikesh

Schiffer Military History
Atglen, PA

Book design by Ian Robertson.

Library of Congress Catalog Number: 2004106993

Printed in China.
ISBN: 0-7643-2097-1

We are interested in hearing from authors with book ideas on related topics.

Published by Schiffer Publishing Ltd.
4880 Lower Valley Road
Atglen, PA 19310
Phone: (610) 593-1777
FAX: (610) 593-2002
E-mail: Info@schifferbooks.com.
Visit our web site at: www.schifferbooks.com
Please write for a free catalog.
This book may be purchased from the publisher.
Please include $3.95 postage.
Try your bookstore first.

In Europe, Schiffer books are distributed by:
Bushwood Books
6 Marksbury Avenue
Kew Gardens
Surrey TW9 4JF
England
Phone: 44 (0) 20 8392-8585
FAX: 44 (0) 20 8392-9876
E-mail: Info@bushwoodbooks.co.uk.
Free postage in the UK. Europe: air mail at cost.
Try your bookstore first.

Contents

Preface

More than half a century has passed since the end of the Pacific War. Interest in the airplanes of that war and the those who flew them remains high. Crashed relics of Japanese aircraft are more and more frequently being brought from the battle areas in attempts to be reconstructed. Other examples that have been warehoused in museums for many years are being restored, and some are newly constructed, bringing their number higher than at any time since the clean-up action following the Pacific War in the late 1940s.

The difficulty in restoring these aircraft is compounded by limited information on the various components and the configurations in which they were installed. This book cannot cover all aspects of specific component types, but if used and studied properly, may prevent the simplest of errors, such as the inadvertent installation of a Japanese Navy instrument in an Army aircraft.

There is a vast assortment of Japanese aircraft components preserved in museums and collections, not only in the United States, but in other parts of the world. Only a small portion of these components are in the hands of people knowledgeable enough on this subject to identify an instrument beyond the basics, and fewer yet could identify the airplane type in which it may have been installed. Much of the intent of this book is to provide a useful reference for the identification of components and matching these to aircraft.

Scale model aircraft builders, particularly those who are interested in Japanese aircraft, will also find this book of value. The author hopes that this collection of photographs is useful for detailing components, and the color information recorded here will be of technical value.

This book follows the publication of three others written by the author on this subject. Two of these were in a series by Monogram Aviation Publications entitled *Japanese Cockpit Interiors*, Close-Up 14 (1976) and Close-Up 15 (1977). More recently, *Japanese Aircraft Interiors* (2000) was published by the same publisher. This book provides additional and more detailed information about the components used in Japanese aircraft.

Those involved in the creation of this edition hope that it will become a basic information source for aircraft restorations, the identification of isolated Japanese aircraft equipment, a useful aid to model builders, and a welcome addition for those that have a genuine interest in the history of aerial warfare.

Robert C. Mikesh
Washington, D.C. 2004

Aichi M6A1 Seiran belonging to the National Air and Space Museum, showing components installed in the rear cockpit.

Acknowledgments

A book of this magnitude and diverse subject cannot be written by just one person. For that reason I am greatly indebted to many people who have willingly helped me gather and consolidate this material, along with conveying knowledge on their areas of expertise. From the very start of this project, Todd A. Pederson has been a prime source of information about Japanese instruments and related components. His private collection of many of these items, which he has closely studied, supports the knowledge he has acquired on this very selective subject.

A close friend and always my mentor on Japanese subjects for the past four decades, Shorzoe Abe, Kobe, Japan, has always been willing to not only translate, but because of his broad technical and historical knowledge in Japanese aviation, adds perspective to early forms of the technical Japanese language that have all but disappeared. He has always been there for me as a guiding hand in the difficult task of conveying Japanese aviation history to the English language reader.

Others have been very helpful in translating the many technical aspects of this subject. Among these are Thomas S. Tomiyama, former U.S. Naval Air Systems Command Engineer. Former Japanese military pilots or air crew members having firsthand knowledge with some of these airplanes that have contributed include Yoshikiyo Ayakawa, Hideaki Inayama, Tsuguo Kawaguchi, Ichiro Naito, and Ryoichi Takashima.

Regardless of the many years spent in the cockpit of larger reciprocating engine aircraft no longer being flown, it is easy to forget mechanical functions that are used in aircraft cockpits for this time period. In addition to my own experience, others helping to recover these important technical aspects that pertained to these components have been Paul Dawson, Kazuhiko Ishizawa, Rob Mawhinney, Robert McLean, Roy Meyers, M. Nakanishi, Howard L. Naslund, T. Watanabe, and Lawrence O. Williams. All with expertise in their own right that have helped document cockpit components that pertain to aircraft engines and accessories, and how they function.

Special thanks go for certain illustrations provided by Giuseppe Picarella of Ifold, England, that describe better than 1,000 words.

Paints and finishes seen in cockpits involve considerable research, and a number of willing participants to help sort out many facts. Among these vital contributors are Donald A. Mader, Kazu Iwasaki, and Vanessa Uneberg, all of whom are with Underwriters Laboratories Inc. (UL); Bob Wade, Chemist with GE; M. Okano of Mitsubishi Heavy Industries Co., Ltd. in Nagoya, Japan; Bruce Ford, Chemist at Australian War Memorial; Cleve H. Hare, Consultant, surface coatings; Tom Hall, James F. Lansdale, James I. Long, and Greg Springer, Japanese aviation historians; Mary Baker and Melvin Wachowiak of the Conservation Analytical Laboratory, Smithsonian Institution.

An author can always use the helping hand of an editor in putting a book together. Once again, Dr. Kevin McCartney, professor at the University of Maine at Presque Isle and long-time aviation enthusiast and friend, has given that help, not only in editing because of his knowledge of aviation, but was able to offer new insights to organization and content. That help is fully appreciated.

Many others have contributed to this work for which I have sought their expertise because of their unique specialty. Some of these are recognized with their credit line on the chapter for which they contributed. Specifically, my thanks for their work goes to Theadore E. Bradstreet, Richard A. Lane, Roy D. Meyers, and Z.I. "John" Szewezyk. To many others as well, that may have inadvertently been overlooked, I deeply appreciate their help.

A final note about photo credits. Credit lines with negative numbers for photographs have been included as much as possible in order to assist other researchers. When a photo source is not given, it can be assumed that the author is the holder. Photos given an *80-G* number are those of the National Archives Record Group and most likely originated from the TAIC effort. Others sources include Signal Corps (*SC*), *Marine*, and *USAF* that are also located at the National Archives.

1

Original Source

Fifty years or more after their manufacture, there still exist an amazing quantity of Japanese components and equipment in private collections and museums. Information on how these items came to be in American hands has all but been lost. Most would have an interesting story to tell of their travels from hand to hand from those war time years to their present locations. This chapter deals with some of the early post-operational history of the various components that will be covered in more detail in the chapters that follow.

With the outbreak of the Pacific War on December 7, 1941, Allied fighting forces had little or no concrete knowledge about Japanese military equipment. Although the air war over China had been in full swing for more than four years, little or no attempt was made by Allied intelligence agencies to learn much about Japanese equipment, tactics, and potential. The reason was simple. Due to severe cuts in military funding following World War I, intelligence units had not been developed for gathering and disseminating this type of information. A secure feeling prevailed of isolation from a conflict taking place halfway around the world, and most concern was focused on Hitler's growing airpower and conquests in Europe.

Once the United States was involved in the war with Japan there was a frantic effort to fill this information gap. Unfortunately, there was very little with which to build an intelligence base for gathering information about Japanese equipment. A meaningful list of Japanese aircraft types was non-existent, much less a way to identify these aircraft when encountered in combat. During this early period, it is thus understandable that every single-engine Japanese fighter was identified as a "*Zero*," and anything else was called a "*Mitsubishi*" or "*Nakajima.*"

It became the primary responsibility of the newly formed Technical Air Intelligence Unit - Southwest Pacific Area to sort out this dilemma. In addition to developing a common means of identifying these aircraft by type, the new intelligence unit obtained information to make drawings and models of these aircraft, as well as acquiring as many photographs of them as possible. Gathering information on performance figures and how the airplanes compared with Allied

Many of the aircraft that revealed Japanese aircraft cockpit technology were found in dismembered condition by the Technical Air Intelligence Units. They analyzed and rebuilt many of these aircraft in an effort to discover and record technical advances that could be useful to the Allies. This is a Mitsubishi Ki-51 Sonia light bomber discovered by members of the 475th Fighter Group. ***Courtesy of Dennis Cooper via John Campbell***

Great effort was often given to the recovery of crashed remains of Japanese aircraft for closer examination. This Kawasaki Ki-48 Lily, ser. no. 534, is being loaded on a barge after recovery at Munda, New Georgia, August 1943. ***Courtesy of J H Lasley.***

aircraft was essential for developing combat tactics with which to counter the Japanese, and in 1942, this was the hardest task of all.

Throughout the scattered Pacific combat zones, fragments of crashed Japanese aircraft and their equipment were being gathered and inspected by the overworked and understaffed military personnel on site at the time. From events such as these developed the field intelligence units. Very little was learned initially from these battered hulks and discarded parts that could set a pattern of production capabilities, design improvement patterns, or air order of battle. The Japanese planned it that way. Nameplates on these aircraft parts were all in Japanese, and translations were often difficult, as few Allied field intelligence personnel read and understood Japanese. The aircraft type identification that was used often confused the issue further. Take for example the early Type 96 Land Based Bomber that eventually was given the Allied code name *Nell*. Its successor became the Betty bomber, which carried the Japanese identifying mark as a Type 1 Land Based Bomber. To most observers, there was no correlation between these two numbers. In time it was learned that "96" pertained to the last two digits of the Japanese calendar year, 2596, which was 1936 in the western calender, indicating the year that the design was accepted as a service aircraft, while the Type 1 Betty was using the last digit of the year 2601 (1941). Intended or not, this type of designation system was more than confusing.

There were many other factors that hindered interpretation. The newly developed radar components carried an identification as a piece of communication equipment. The *Shi* found in certain designations meant *experimental*, yet this marking was often carried on service equipment all through its quantity production cycle. Serial numbers on aircraft were frequently designed to be misleading. While block numbers were used, the prefix number or numbers scrambled the sequencing, a system which has been fully understood only in recent years.

The retrieval of an airplane from the jungles of the South Pacific was a laborious task. Here local natives recover the forward fuselage of this G4M1 twin-engined Betty, serial no. 1365, near Munda, New Georgia Island, February 10, 1944. ***Courtesy of J L Lasley.***

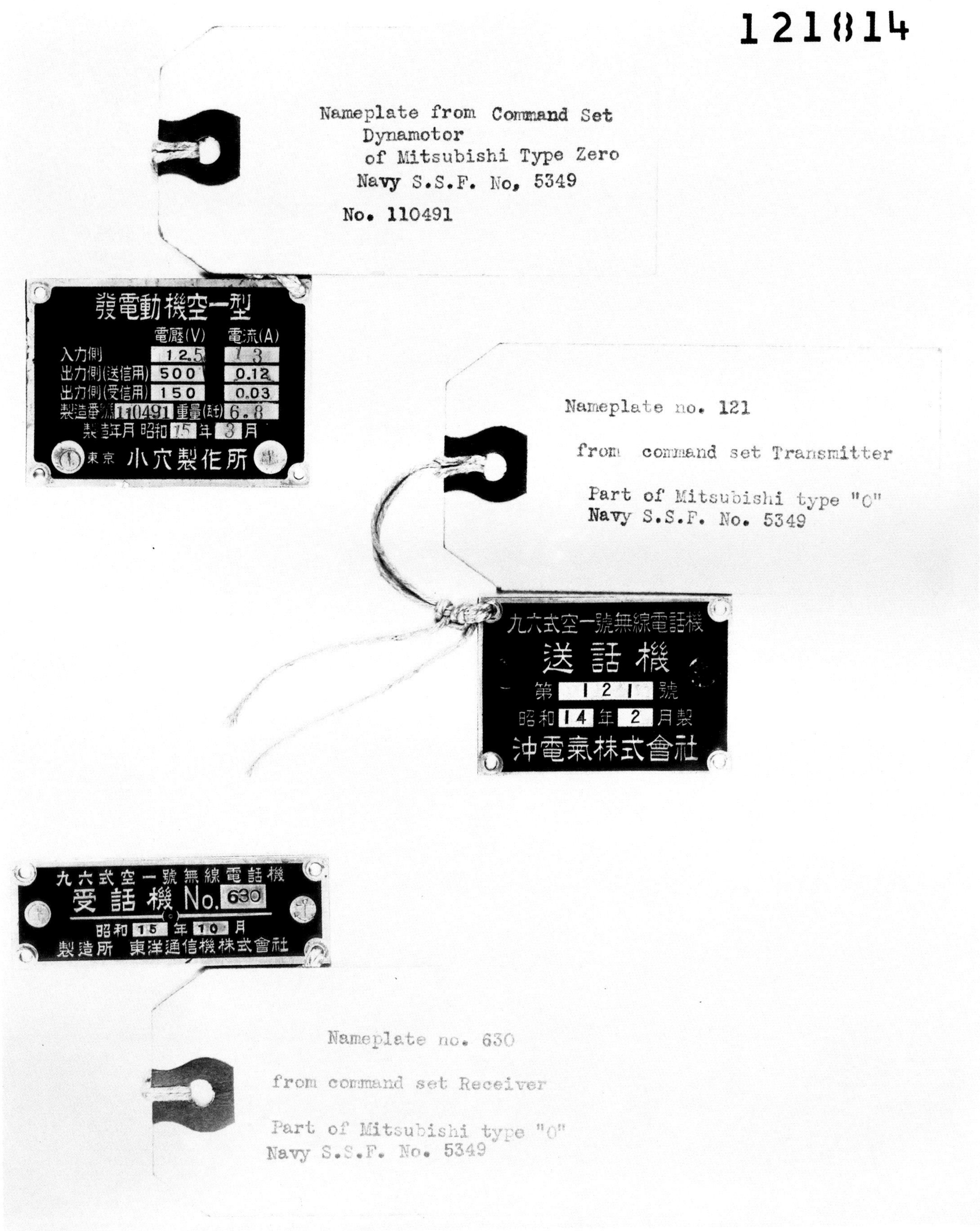

These are examples of identification placards retrieved from downed Japanese aircraft, tagged, and their data recorded for comparison with similar information. This would reveal production locations, and hopefully production rates by comparing serial numbers with production dates. These placards were gathered from a Zero fighter. For aircraft serial numbers, the Japanese developed a code system that generally varied with different types of aircraft to confuse intelligence gatherers.

When personnel of the Tainan *Kokutai* were ordered to return to Japan, less their battle-weary equipment, prime examples of A6M3 Hamps were left behind at Buna airstrip, New Guinea. Two of the best airframes were laboriously disassembled and, along with three Sakae 21 engines, were carried and dragged through the jungle and mud to the beach and a waiting barge. These parts provided the intelligence technicians at Eagle Farm, Brisbane, their first real opportunity to produce a flying Japanese fighter for evaluation. ***Courtesy of Mathew C Gac.***

The person given direct authorization for the project was General George C. Kenny, commander of Allied Air Forces in the Southwest Pacific Area (SWPA). Kenny's personal experience, technical qualifications, and broad interests gained in early assignments at Wright Field in Dayton, Ohio, before the war, sparked his attention and realization of Allied shortcomings in the knowledge of their Pacific enemy. It was sadly apparent to Kenny that there was no formalized air intelligence gathering system in existence within that theater. The Allies were, in fact, profoundly ignorant concerning Japan's aviation capabilities.

To assimilate this information, Technical Air Intelligence Unit (TAIU) was organized from a small group of intelligence personnel that formed in Melbourne, Australia, and soon moved to Brisbane. As part of the 5th Air Force and Allied Air Forces in early 1943, the TAIU served as a focal point for intelligence information. Its staff headquarters was in the AMP Building in Brisbane, while the technical examination, reconstruction, and simulated combat trials with captured Japanese equipment were located at a newly developed airfield near Brisbane, known as Eagle Farms. A team of U.S. Army Air Forces technicians, aided by the U.S. Navy, RAF, and Australian military forces, set up shop in a remote location, inside a specially constructed and equipped hangar. This test facility, simply referred to as "Hangar 7," was a highly restricted area, and soon was generating a great amount of interest about the Japanese equipment being evaluated there.

It was at Eagle Farms that downed or captured Japanese aircraft and equipment were brought to be examined under conditions far better than the remote jungle areas from which most of this equipment was originally acquired. This effort in the theater where the fighting was taking place was made even more pivotal by the fact that captured equipment already sent to the United States early in the war was being set aside and not being studied with the intent of gathering and correlating intelligence information. Apparently, the nation was more involved with increasing war production and the training of fighting personnel than it was in studying details of captured equipment, and this lack of interest showed in our lack of information about the enemy.

Captain Frank T. McCoy, Jr. (later Major General, USAF), was in charge of this Technical Air Intelligence Unit - SWPA, and was most helpful in the development of this account about the TAIU. He, along with two of his intelligence specialists, T/Sgt Francis "Fran" Williams from Rio Rancho, New Mexico, and Corporal Joseph Grattan from Pittsburgh, Pennsylvania, had come from Jackson Army Air Base, Mississippi, to Ballarat, Australia, as part of the

Souveniring by American G.I.s was the greatest threat to the TAIU teams in removing vital intelligence information from aircraft before the information could be recorded for technical purposes. Here a G.I. *gingerly* removes the kanji tail markings from this Ki-84 Frank for a souvenir. Obviously, TAIU teams had already been here, since they had taped the aircraft as scale for measuring from photos they would have taken.

Pilot's perspective when entering the cockpit of a Kawanishi N1K2-Ja George. Vertical entry to cockpits allowed equipment along each side to be unhindered by entry passage.

Kawanashi N1K2-Ja *Shinden Kai* George Navy fighter.

The location of this battered Mitsubishi G3M2 Nell is unknown, but the picture was most likely taken at Atsugi AB, near Tokyo, when newly occupied. The orange color indicated test aircraft or trainer, although an unusual camouflage pattern suppressed the brilliant color. ***Courtesy of Jacques Young via Jeff Ethell Collection.***

38th Bombardment Group. As the obvious importance of intelligence gathering of Japanese equipment grew, McCoy "requisitioned" other technicians from his former and other units to create a team of trained aircraft specialists who were tasked to retrieve aircraft components and turn them into flyable examples to be flight-tested against Allied counterparts.

It was from the very early days of the formation of this unit at Victoria Barracks, in Melbourne, that a plan was set in motion for the capture and retrieval of Japanese equipment in order to discover their innermost secrets. After sightings were reported of downed aircraft that appeared to be in reasonable condition, a Technical Air Intelligence officer and a small team of men would be dispatched to the site. The success of their mission was heavily dependent upon the assistance of those already in the area. This help came from a broad range of sources that included U.S. forces and their Allies that were occupying the areas, Coast Watchers, village headmen, and local police, as well as natives, many of whom spoke in "pidgin English" as taught to them by missionaries. The nearer to the fighting area, the greater the cooperation in obtaining captured equipment.

One of the best aids in amassing the valuable aircraft parts came through the assistance of native work crews, who have never been given the credit they deserve. They were the ones that often spotted the wreckage, and then assisted in dismantling and transporting the enemy aircraft from often remote areas to the tidewater, hauling them through swamps, or floating them down rivers on rafts. Although there was a wide variation among the islands, most natives were intelligent, good workers, and extremely loyal. They became skillful with wrench, chisel, and hammer, but showed little interest

This Mitsubishi Ki-67 Peggy has been stripped of every removable part before Allied forces occupied this unidentified Southwest Pacific location. ***Courtesy of Jim Hoton via John Campbell.***

Tandum seating in this Ki-55 was uniquely different. Cutout at right of front cockpit instrument panel was space for machine gun, while rear instrument panel appears inverted. ***Courtesy of Shorzoe Abe*****.**

Tachikawa Ki-55 Ida Army trainer depicted by model aircraft. ***Courtesy of Shori Tanaka.***

Japanese aircraft were also found in China, many of them confiscated by the Chinese for use in their post-war air force. Shown here is a battered but complete Kawasaki Ki-48 Lily and a Nakajima Ki-84 Frank. Note Chinese insignia on the side of Lily. *Courtesy of Larry Davis.*

This view inside Hangar 7 shows the stockpile of Japanese aircraft parts. In the foreground are center sections for three Oscars and one Sonia (right), with four cockpit areas of Zeros and a Betty tail in the background. Note the stacked propellers in between. *Courtesy of Moring P. Clark.*

in the planes themselves, unless as a source of some material that might be used to make jewelry, a pipe cut from an oil line, or even a hat fashioned from airplane skin. Since Japanese pilots did not always pick convenient beaches to crash on, TAIU officers sometimes labored for weeks to get the wrecks out of the dense jungles. Without the assistance of these jungle-wise natives, they never would have been able to make their way through unhealthy and often uncharted territory teeming with disease carrying insects

Replacements fresh from the U.S. for the teams sometimes underestimated the natives' intelligence, however. One story has it that a newly arrived G.I. asked a bare-chested native, complete with spear; "how you like'm ride big bird," while airborne on a PBY trip to assist at a crash site. "Quite an interesting flight" was the reply with a trace of British accent.

During the early stages of the war most natives were terrified of the big "Baloose," usually a battle weary C-47, and would lay face down on the floor during most of the flight, yet they flew anyway.

In another example of native cooperation, one Navy Corsair pilot swore that he had shot down a Focke-Wulf Fw 190 over Santa Isabel Island in the Solomons. The plane crashed six miles inland in thick jungle, atop a 1,500 ft hill. To get to it and bring out the engine would have taken three weeks, the American technicians estimated. However, since it was later determined that the plane was the newly introduced Judy dive-bomber with an in-line engine, it was especially desired for analysis. A crew of natives, who were obviously familiar with the area, worked for a shilling (16 cents then) a day, hewing a trail with machetes to the crash spot, and carried the engine out of the jungle on a bark-woven cradle—in four hours.

During their stay in Australia, the Technical Air Intelligence Unit - SWPA had not been exposed to new Japanese aircraft that were expected to have been introduced in that theater. They found, instead, that both the Japanese Army and Navy were sending in improved models of previously existing types. There were reports of new aircraft being developed, and some had been seen, but none had fallen into Allied hands.

Following the Hollandia Campaign on the northern portion of New Guinea, the Eagle Farms TAI Unit received sudden orders in June 1944 to return to the United States and set up their operations at Anacostia Naval Air Station in Washington, D.C. This was a result of a decision influenced by McCoy, now a Lt. Col., for consolidating all Technical Air Intelligence activities, captured Japanese documents, and nameplate analysis closer to the Pentagon. To these technicians this was welcome news, but a bit bewildering, since their intelligence gathering work seemed essential in its being accomplished near the war zone, instead of in the capital of the United States. Nevertheless, the Navy dispatched a "Victory Ship" to Brisbane for the unit's exclusive use in transporting the men and their equipment to San Francisco, followed by a private train for the trip to Washington. The two and a half year stay in Australia for this major intelligence gathering unit had come to an end.

Australian and American technicians work side by side on this center section of an Oscar in Hangar 7. Not only is the airframe being prepared to be flown, but in the process attention is given to any details that might be of technical interest. ***Courtesy of Moring P. Clark.***

This restored cockpit of the sole surviving Ki-100 becomes quite colorful when faded colors are freshly painted. ***Courtesy of Giuseppe Picarella.***

Kawasaki Ki-100 Army fighter ***Courtesy of Giuseppe Picarella.***

They did not leave the war zone, however, without an intelligence collecting capability. For the war in the Pacific area, all intelligence sources were governed by the Technical Air Intelligence Center at Anacostia Naval Air Station. Under this Center were all the field TAI units in the Southwest Pacific Area (SWPA), Pacific Ocean Area (POA), Southeast Asia (SEA), China, and India. These units would recover and make the first-phase examination of captured air equipment on the spot, and then see to it that these aircraft and material were sent back to the TAIC as soon as the theater had completed study for immediate use.

These field TAI Units were headed by a trained TAI officer and included aviation mechanics, photographers, radiomen, ordnance men and, oftentimes, translators, depending on conditions. They were equipped to travel into any kind of territory and make use of whatever transportation was available. In isolated areas, such as many parts of China and India, field TAI Units often consisted of only one officer and one enlisted man, depending almost exclusively on local help. In other places, where many aircraft were located on one newly occupied enemy air base, for instance, large numbers of TAI personnel participated in the recovery of enemy equipment, often supplemented by men and equipment of combat units campaigning in the area.

Technicians at Anacostia busied themselves rebuilding Japanese aircraft brought with them for flight evaluation. In order to make them airworthy, every major assembly was disassembled to inspect and record the engineering features. While being reassembled, each operating component was checked to make sure that it was serviceable for flight. In most cases technical manuals were not available, much less translated into English, so everything had to be worked out on a trial and error basis. While this was challenging, it was not too complicated, for the state of the art was quite similar among all airplane manufacturers of that time period.

In most cases, Japanese flight instruments were replaced by American types for those airplanes intended for flight. It was often necessary to retain the original engine instruments, however, since they were designed and calibrated for the function they served. In all cases, each instrument was identified in English to replace the Japanese labels. American radio equipment and entirely new oxygen systems also replaced Japanese types.

Of the aircraft known to have been reworked at NAS Anacostia and flown for these evaluation flights were the Zeros that arrived from Saipan, and two Oscars; one a Model 2, and the other a very late-war Model 3. Despite engine problems the Tony was flown, along with an advanced version of the Ki-46 III Dinah. One reference was made to a Rufe at Anacostia, but no pictures or reports support this. The oldest of the group was the Kate, which was made airworthy and appeared in several airborne photographs in Japanese marking scheme for recognition training purposes.

This Ki-43-I Oscar is at rest on the run-up area at Townsville in March 1944 after rebuilding by TAIU at Eagle Farm. Development of Oscar as an Army fighter was a year or two behind that of the Zero, yet both had the same engines with only minor differences. Stripped of all camouflage and with the addition of American insignia, this Oscar is readied for flight evaluation. ***Courtesy of Australian War Memorial.***

With the occupation of Hollandia, on the northern coast of New Guinea, the Allies recovered a number of Japanese aircraft considered to be salvageable. This Ki-46-II Dinah was the prize. With the aid of 49th Fighter Group mechanics stationed there it was made flyable within several months. Shown here on September 14, 1944. The ranking officer pilots were privileged to fly this aircraft before it was shipped to the United States around January 1945. *Courtesy of Francis M Williams*.

Irving's instrument panel was unique in that it had a cutout in the middle for a downward mounted gunsight for firing the ventral oblique cannon. This feature was deleted when radar was added, but the opening remained.

Nakajima J1N1-S *Gekko*, Irving night fighter.

When the war ended so did the emphasis for further evaluation of these aircraft. Japanese aircraft of many different types had already arrived from Japan in flying condition and been sent to other flight test facilities, but even that program did not meet with great popularity. In time, most of these aircraft were pushed out of sight and soon disappeared in the bulldozed heaps of scrap that brought an end to so many war planes in the late 1940s.

The volume of Japanese aircraft components gathered throughout the Pacific and Occupied Japan was enormous. To solve the storage problem for equipment no longer having to be evaluated, the former Douglas Aircraft assembly plant that was government-owned at O'Hare Field (now Chicago International) was selected for warehousing this equipment. Flight instruments, radios, bombsights, and cameras of every description were in large wooden boxes stacked in hangar-like buildings. Adjacent buildings and the outside ramp area housed examples of Japanese and German aircraft, along with American types. Name it, and it could probably be found in this huge stockpile.

In 1949 this became the foundation for the National Air Museum, later the National Air and Space Museum. Nearly the entire collection was transferred to the new museum and accepted in place, as there was no place else for it to be stored. Then came the Korean War. Almost overnight these government buildings were given new meaning, for it was planned that they would be used for the assembly of Fairchild C-119 Flying Boxcars. The museum had to vacate the buildings—immediately!

To begin the process, it was apparent that the amount of excess property was out of proportion to what could or should be retained by the museum. The minimum staff, not all that knowledgeable of German and Japanese marked equipment, along with Air Force personnel assigned to help, began the process of reducing the stored materials to something more manageable. This was done as well as could be expected under these trying circumstances. Most of the collection was deemed excess and went to scrap yards—everything from nearly flyable airplanes and engines, to oxygen masks, bombsights, and instruments. One example was the Nakajima Ki-84

When the Allies advanced to Clark Field in the Philippines in February 1945 they found a large assortment of Japanese Army and Navy aircraft dispersed on and around the field. One of nearly every type of aircraft found was made airworthy and evaluated. Since the field was occupied before the fighting on Luzon ended, the red hinomaru of the Japanese insignia were painted blue, so that when spotted from the air by Allied fighter pilots, they would not be mistakenly strafed.

Above: The instrument panel for Oscar was compact in order for two machine guns to be mounted on either side of the panel. Below: Nakajima Ki-43 *Hayabusa*, Oscar Army fighter. *Courtesy Albine Figher Collection.*

The objective of retrieving Japanese aircraft was primarily to evaluate them against Allied counterparts. Shown here is a Ki-84 Frank leading a Spitfire, Hellcat, and Mustang during flight evaluations in the Philippines. ***80-G-193335***

Frank that was rejected and bought from a nearby scrap yard by Ed Maloney of Planes of Fame, Chino, California. After this airplane was made airworthy, the fighter was eventually purchased by a Japanese businessman and returned to its home land. There were many similar accounts of the acquisition of this excess material.

Collectors and war surplus buyers soon learned of this material being disposed of in the late 1940s. German and Japanese equipment came on to the open market, while much more went to the smelter. With the relatively few exceptions from servicemen who brought home flight instruments or similar war trophies, this surplus from Chicago became the origin of most equipment found in the hands of collectors and museums today. What a fascinating and unique story each of these captured pieces of enemy equipment could tell of its circuitous route from Japanese manufacturer, into the war zones, through many hands and locations over these past 50-plus years to their present owners.

Guns and charging handles on either side of the instrument panel can be seen in this Zeke 52 cockpit. Artificial lighting distorted exterior green color to a false blue. Both views are of the NASM A6M5.

Mitsubishi A6M5 Zero-*sen*, Navy fighter.

The Sakae 21 engine of this Zeke 52 coughs to life after having been made air worthy by the TAIC at NAS Anacostia. This was one of approximately a dozen Zeros captured at Saipan in June 1944 and brought to the U.S. to rebuild a few for evaluation. This airplane now is on display at the National Air and Space Museum.

At Park Ridge, Illinois, on what is now Chicago International, two large warehouses were filled with airplanes and equipment, both U.S. and foreign. Seen here in Building T-6 is a Ju 388 on the right, and Kugisho P1Y1 Frances on the left, with rows of others down the center. Both aircraft still exist in the National Air and Space Museum collection.

Seiran's **crew of two operated the Navy Type 3 Gun/Bombsight in front of pilot, and the dive angle computing mechanism Navy Type 1 Control Unit Gyro Stabilized Inclinometer at top right in front of rear seat crew member.** ***Courtesy Thomas S. Momiyama.***

Aichi M6A1 ***Seiran*****, submarine based bomber, newly restored by the National Air and Space Museum.**

In addition to airplanes, the collection, while in Chicago, consisted of every conceivable piece of foreign aerial equipment, often duplicated many times packed in these boxes. The vast quantity caused much of this to be sold as scrap, allowing collectors to acquire this equipment.

Priceless optics, such as these Japanese Army Type 88 Bombsights, are tossed for scrap at some unknown location. *SC 216780*

For the one-way flight of the *Ohka* 22, this cockpit had little more than the bare essentials in instrumentation. Note crudeness of replicated wooden throttle handle. This sole survivor is in the National Air and Space Museum.

Kugisho *Ohka* 22, kamikaze bomber with author present. Note Venturi tube on right side by cockpit.

The spaciousness of the four-engine Emily flying boat provided room for this radio compartment, and to the rear was a flight engineer's station. Navigator's plotting table was to the port side.

Kawanishi H8K2, Emily Navy Type-2 flying boat.

As crude as this Ki-115 *Tsurugi* aircraft might be for its one way *kamikaze* mission, the two side by side buttons under the magneto switch ignited left and right JATO thrusters for its final high speed plunge into the target.

Nakajima Ki-115 *Tsurugi*, Army *kamikaze* bomber.

2

Aircraft Instruments

Instrument Identification

An important aspect intended for this book is to serve as an aid in identifying Japanese instruments that have had their labels separated, or as a guide when the label cannot be understood or read. The system within this book is designed with the assumption that the function of the instrument can be determined on sight, since appearance in function is basically the same regardless of nationality.

The index at the end of this book separates Army and Navy listings of instruments and are grouped within these two military branches by function. Therefore, if a tachometer is to be identified, and the branch of service has not been determined, one must search both listings under tachometer, and refer to this section for further description and possible photograph. At least one of these lists should contain the instrument in question.

The branch of service (Army or Navy) can sometimes be identified by an anchor stamped into the placard for Navy, and a five pointed star inscribed in a circle for Army.

There are other ways that a collector of these instruments or managers of collections should be able to interpret these identifying placards on instruments. First there must be an understanding of the way that instruments are labeled.

Generally, one will find the Japanese military nomenclatures to follow this pattern:

Navy Type 0 Aircraft Compass Model 2 Kai-1

For listings of instruments within this book, this format has been altered so that they can be placed in functional group listings. For example, when using this book, a search would be made for "compass," not Navy or even Type, since that designation applied to more than just compasses. Therefore, for simplicity, the author has taken the liberty to list the same instrument as follows:

Compass Type 0 Model 2 Kai-1

eliminating the term Navy and Aircraft unless the description is used within the text portion in which the normal form of terminology will generally be used. One must remember that these service names as a prefix are an official part of most nomenclatures. While commas would seem appropriate, for brevity they are not used. There are times that the detailed identity of Model, Type, etc. are not given or known, and are left out of the identification. To provide continuity when reviewing instrument lists—such as in the index—the basic word description, such as "compass," will always be consistently at the left. One must understand that when searching for kanji characters that describe the component, this sequencing has been rearranged in this book.

In translating these nomenclatures, the two kanji of "*Koku*" (航空) often appear. In translation dictionaries, it means "aviation," "aircraft," "flying machines," and others. Since "*koku*" is used to differentiate usage from such things as "marine" (compass), the most suited identifying term for which it is intended would be "aircraft."

Typical Japanese instrument panel presentation, with instruments of same dimension as U.S. types, and each labled in Japanese.

To be able to read Japanese is not essential in finding the Type and Model from instrument placards. By comparing the basic kanji contained in this section for identity, one can determine that this compass label reads:
Type 0 Aircraft Compass Model 2 Kai 1
Number: (Stamped in) Mk. (Navy stamp)
Manufactured: Showa (year) Month
Tokyo Koku Keiki K.K.
Courtesy of Thomas Davidson.

Some instruments that are relatively basic are without Type or Model numbers, yet have this identity elsewhere. The placard on this instrument simply reads:
Airspeed Indicator
Number: (Stamped in) Mk. (Navy stamp)
Manufactured: Showa (year) Month
K.K. *Tanaka Keiki Seisakusho*
Courtesy of Thomas Davidson.

The anchor immediately identifies this as a Navy instrument, and therefore is not part of the lettered nomenclature. Here is another example of Type and Model missing, and therefore the face must be compared with others in order to determine the full nomenclature.
Rate of Climb
Number: (Stamped in)
Manufactured: Showa (year) Month
K.K. *Tanaka Keiki Seisakusho*
0 Point Adjustment.
Delivered: Showa 15 (1940) April
To Constant Temp. Tank. Air Vent
Courtesy of Thomas Davidson.

There are levels of identity within these nomenclatures for which terms like Type, Mark, Model, etc. are used, and therefore warrant further explanation. These will be covered with the most basic term first, which also follows the format in which nomenclatures are normally written in Japanese usages. It is rare that all the terms are used in any one nomenclature, and there are known instances where these vary in sequencing. A typical Navy fully detailed instrument description may be written as follows:

(Kanji) 海軍九二式二號航空羅針儀一型改一:
(Romanji): **Kaigun 92 Shiki 2 Go Koku Rashingi 1 Gata Kai 1**
(Translation with sequencing arranged English style.)
(English)**: Navy Type 92 Mark 2 Aircraft Compass Model 1 Kai 1**

Navy (Terminology Explanations):

Kaigun (Navy) **海軍**: This identification will always be first in the fully detailed nomenclature.

Shiki (Type) **式**: The word "*shiki*" has several translations, all similar to and including "type." Since early issues of *Jane's All The World Aircraft* began using the word "Type" for "*Shiki*," and this has become the accepted standard translation.

The Navy Type numbers used relate to the year of acceptance as a military object. Examples are Type 98, having been accepted in the Japanese year 2598 (1938), and Type 3, year 2603 (1943).

Go (Mark) **號**: Numerous usages of "*Go*" are found in Japanese terms. The most common usage identifies an object within a series, like the 4th, not number 4 as a serial number. Since more than one item (like compass) may be accepted in a given year (expressed as Type number), the usage of different "Mark" numbers can make this distinction—but is not always the case.

Name of object (description): This is often more explanatory than merely "compass." Included might be the term "pilot" or "navigation," denoting usage, whereas English terminology generally describes function such as "magnetic," "remote," or "fluxgate," as in the case with compasses.

Kata (Model) **型**: "*Gata*" is an alternate spelling and pronunciation. Translations vary, however, but since Jane's wartime usage established "Model" for *Kata*, this has remained the standard. Uses of "Model" can vary depending upon the object category.

Kai (Modification) **改** followed by a sequential number: Self explanatory.

A typical Army fully detailed instrument description may be written as follows:

(Kanji): 陸軍一式二號人工水平儀一型甲
(Romanji): **Rikugun 1 Shiki 2 Go Jinko Suiheigi 1 Gata Ko**
(Literal translation line with words sequenced as above.)
(English)**: Army Type 1 Mark 2 Artificial Horizon Model 1 Ko**

From the above example, it becomes apparent why in translation numbers follow the identifier to which they apply, i.e., Model 1, not 1 Model.

Army (Terminology Explanations):

Rikugun (Army) **陸軍**: This identification will normally be first in the fully detailed nomenclature.

Shiki (Type) **式**: Same interpretation as with Navy.

Go (Mark) **號**: Same interpretation as with Navy, but is seldom used with Army instruments.

Name of object (description): Same interpretation as with Navy.

Kata (Model) **型**: Same interpretation as with Navy.

In the example above this is *Ko*, which in English could be intended as "A." This is a suffix to Model. Translations often use A, B, C, etc. for the characters of *Ko* (甲), *Otsu* (乙), *Hei* (丙) *Tei* (丁), *Bo* (戊), etc. Either system is appropriate in translation from Japanese sources, as long as there is consistency within groups of items. The letter alphabetic is more generally used with aircraft nomenclature, while the Japanese form is more often used with components.

Numbers found in nomenclatures are as follows:

零	一	二	三	四	五	六	七	八	九	十	百	千
0	1	2	3	4	5	6	7	8	9	10	100	1000

Manufacturers of Instruments

When an instrument is identified by nomenclature, it is often appropriate to have the name of the manufacturer. To aid in this translation, the major manufacturers of wartime instruments are listed below. Their kanji is given with these names for comparison to the instruments. These names are generally written without spaces, but for this guide, the words have been separated so as to be matched to the translation.

Aichi Kokuki	愛知 航空機
Dai Ni Kaigun Kokusho Heikibu	第二 海軍 航空廠 兵器部
Fuji Koku Keiki	富士 航空 計器
Hitachi Seisakusho	日立 製作所
Kayaba Keiki Seisakusho	萱場 計器 製作所
Mitsubishi Denki	三菱 電機
Nikko (abv. for Nippon Kogaku Kogyo K.K.)	日光
Nippon Seimitsu Denki	日本 精密 電機
Nippon Kogaku Kogyo K.K.	日本 光学 工業 株式会社
Shimazu Seisakusho	島津 製作所
Shinagawa Seisakusho	品川 製作所
Tanaka Keiki Seisakusho	田中 計器 製作所
Toko (abv. for Tokyo Kogaku Kikai)	東光
Tokyo Koku Keiki	東京 航空 計器
Tokyo Kogaku Kikai K.K.	東京 光学 機械 株式会社
Tokyo Keiki Seisakusho	東京 計器 製作所
Yanagi* Seisakusho	柳 製作所
Yokogawa Denki Seisakusho	横河 電機 製作所

* Kanji used by this company may be different.

Army Type 96 Air Speed Indicator. ***80-G-192550.***

Identification to Aircraft

When examining an instrument, it is important to learn what type of aircraft it could have served. For that purpose, lists of instruments in Appendix A (Army) and Appendix B (Navy) have been developed that identify the type of aircraft they are known to have served, and those in which they could have been installed.

With more than half a century of time that has passed, it is impractical to assume that a complete list of instruments can be developed. It is believed, however, that the listings contained in these two appendices are as lengthy as any other now in published form. It is hoped that this will form the baseline to which others can be added.

Selected Instruments and Functions

To the aviation literate person, the purpose or function of an aircraft instrument is generally visually identifiable. With the passing of time, however, and newer instruments coming into use, the function of the earlier types mentioned in this book for their time period are easily forgotten and perhaps not well recorded. As an example, to recognize the calibration of their scale, if not marked, or to understand how that unit of measurement is used can become a forgotten issue. For these reasons certain instruments have been selected for this section to make note of their usage and calibration else they be lost.

To find the answers to some of these seemingly logical questions has already become difficult. At the time that certain controls for reciprocating engines were used, or flight instruments were in common usage 50-plus years ago, the information these instruments conveyed was everyday knowledge. Following the war, Japanese aviation came to a standstill for nearly 10 years. When it was reborn, foreign equipment was put in service, and the jet age also changed many things. The purpose of some cockpit controls and functions of instruments was out of mind and soon forgotten. It is hoped that this section will bring back, for the record, some of these aspects of the internal workings of these controls and selected instruments that were once so common.

Airspeed Indicator

The airspeed indicator shown here would have been placed in Japanese service in 1936 for the Army because of its Type number. This instrument is graduated in kilometers per hour, as this was the Army's standard measurement of speed and distance. Its pointer makes one and three quarters revolution to reach its maximum reading of 600 km/hr, which is 373 mph, or 323 kt.

This Navy airspeed indicator (upper left, p 35) placed in service in 1942 shows a maximum reading of 300 knots, and thus different than the unit of measurement of the Army, which used km/hr. The same difference occurs in the U.S. military, since the Army uses miles per hour and the Navy uses knots for speed measurement. Conversions of maximum speed for this instrument of 300 kt are 345 mph and 555 km/hr.

Navy Air Speed Indicator Model 2 taken from Judy. ***80-G-191883.***

Army Type 98 Turn & Bank Indicator manufactured by ***Tokyo Koku Keiki*****. 80- G-191879.**

Turn and Bank Indicator

The turn needle is a part of this instrument that is beginning to fade from usage, yet is still found in older aircraft. To most pilots that can now rely upon gyro stabilized instruments, the turn needle is known only to indicate the direction of turn. The primary purpose of this turn needle was for making time-turns in the days of early instrument flying and the WW II time period, when gyro horizon instruments were sometimes unreliable. These procedures often depended upon precise 180 degree turns, or one of several procedure turns, the simplest being 90 degrees to the right and 270 degrees to the left (for propeller driven aircraft) in order to reverse course on the same flight path. By establishing the amount of turn that would deflect the turn needle one needle width, the turn rate was three degrees per second, a full 360 degree turn in two minutes, or portions thereof. Those were the times that the chronometer or clock on the instrument panel was a critical flight instrument for timing turns.

The ball in the liquid-filled tube was used to correct for skid or over-banking, causing an uncoordinated turn. These acted much like a carpenter's level, but in this case, affected by the force of the turn. Keeping the ball centered assured a well coordinated turn. And too, if the ball is not centered, the guns will not fire at the point the aircraft is aimed.

Army Turn & Bank Restricter Valve and Gauge manufactured by ***Tanaka Keiki Seisakusho*****. 80-G-192295.**

Vacuum Gauge

Although this instrument translates to Suction Gauge, the accepted English terminology has been Vacuum Pressure Gauge. It was not until after World War II that gyro instruments were electrically driven. Up until that time, Venturi Tubes and engine-driven suction pumps drew air out of gyro-operated instrument cases, and incoming air was directed through a nozzle to the vanes of the gyro, causing it to spin. The maintenance of a prescribed vacuum (lowered pressure, actually) to control this jet of air had much to do with the accuracy of the instrument. An adjustment knob on the instrument regulated the pressure shown on the dial. Because of the critical importance of the Turn and Bank Instrument when flying in weather, some engines had two pumps and a selector switch in the cockpit, or in multi-engine aircraft, each engine had a pump and a selector switch in the cockpit. This switch or valve lever often contained a drain-line position, as well.

Vacuum or low-pressure source for this system was first provided by Venturi Tubes, a double funnel-like device attached to the outside of the fuselage. Air passing through the throat of this device caused low pressure. A tube from this point on the Venturi to the suction end of the instrument case provided a source of air for spinning the instrument gyro, usually limited to the basic Turn & Bank Instrument. A photo of a Venturi Tube is shown on the Ohka 22 on page 28.

Gyro instruments that used this low pressure air system were the Turn & Bank Indicator, Directional Gyro, and Artificial Horizon. Of course, the latter two instruments were major components in Autopilots. If the pilot noted low pressure that could not be corrected, he had to elect turning off one of the instruments to increase the pressure. The Turn Needle was retained to the last.

Army Type 95 Rate of Climb Indicator Model 2 manufactured by *Tokyo Koku Keiki*. 80-G-191876.

Rate of Climb Indicator

The Rate of Climb instrument is one of three instruments connected to the pitot static air system, the other two being the airspeed indicator and altimeter. Graduated in meters per second in climb or descent, the 5 graduation mark represents 500 meters (1,640 feet) per minute. The needle deflection for rate was the same for American and Japanese instruments, only the graduations were changed. At the time this instrument was heavily relied upon; normal rate of ascent and descent on instrument flight were 500 ft/min as an American instrument flight standard.

Army Type 97 Sensitive Altimeter manufactured by *Tanaka Keiki Seisakusho*. 80-G-191832.

Altimeter

This type of altimeter is a straightforward instrument having two pointers for very accurate reading. One full sweep of the larger pointer would place the small hand at 1, indicating 1,000 meters (3281 ft) of altitude. A full revolution of the small hand would take it to 10,000 meters (32,810 ft.). The vernier scale in the window at the bottom can be set by the thumb knob below. This vernier is

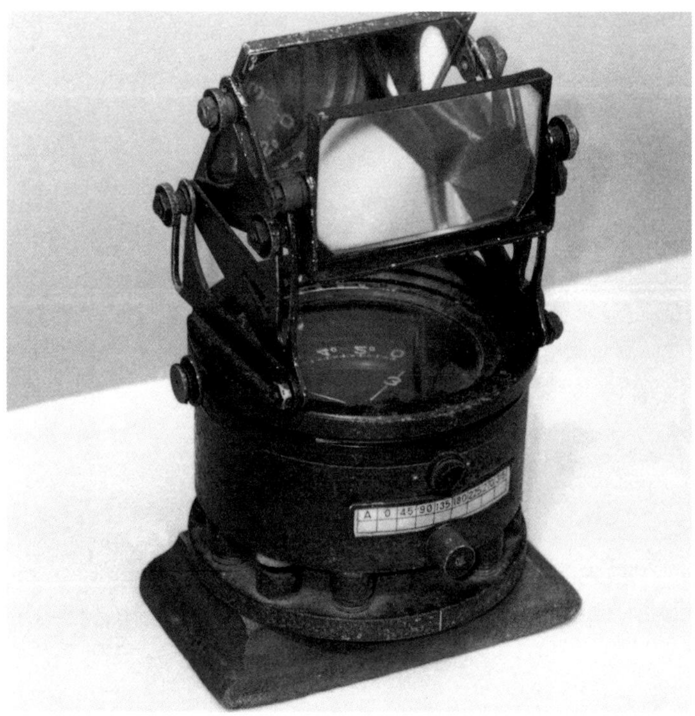

Navy Navigators Reflector Compass Model 1

graduated in millimeters of mercury (mm Hg), while the American instruments were calibrated in inches of mercury as the measurement for barometric pressure. This dial is known as the Kollsman window on American altimeters. Since atmospheric pressure changes with weather conditions, this altimeter can be set for the current pressure while in flight, transmitted from a ground station, without having to set it at ground elevation while at rest.

Air Navigation Reflector Compass

Precision navigators' compasses (bottom right, p 36) are normally larger by comparison to pilot instrument panel varieties. These are vertical reading instruments that are viewed from the top. In small aircraft cockpits these become difficult to read. By attaching a reflective mirror capability, along with reversed images, the compass reading becomes horizontal.

Vertical view of Reflector Compass with glass reflectors and apparatus removed. Small outer scales are adjustments for magnetic variation and deviation. The 100 degree inner scale is normally more centered toward the back (top) of the case.

Shown here is a Navy Type 3 Reflector Compass Model 1 in a partially disassembled condition. This is a down-looking view at the compass card and instrument face. The pointer that swings with magnetic influence consists of four arms permanently joined with 100-degree separations, except that the fourth quadrant is 60 degrees to complete the 360 degrees of the compass. At the back of the case are fixed graduations of 100 degrees. All these images are reversed, since they are to be read through a mirror.

When the compass is to be read, only one pointer at a time will be within the stationary 100 degree scale. That numbered pointer identifies the hundred degree heading, and the scale to which it points identifies the tens degree reading. To read the vertical image pictured here (without the aid of mirror reversal), the heading of the aircraft would be 116 degrees east of north.

The operator's view of the Navy Navigator's Reflector Compass Model 1. The reading of 133 degrees is reflected into the glass reflector.

What the user will see when looking horizontally at the compass is shown in this photograph, which is influenced with UV light to enhance the image resolution. The 100 degree scale and associated one pointer is projected upward to the tilted glass for horizontal reflection to the viewer.

Army Type 98 Artificial Horizon manufactured by *Tokyo Koku Keiki*. 80-G-191905.

Artificial Horizon

Selecting the proper term for this instrument (bottom left, p 37) to use throughout this book became a problem. Wartime terminology describing this instrument consisted of Artificial Horizon, Attitude Indicator, Flight Indicator, Gyro Horizon, and others. This latter term, however, was actually the gyro stabilized horizon bar in the window of the instrument that was more correctly identified as the Attitude Indicator. Literal Japanese translation, however, is Artificial Horizon for the Army, while the instrument label for Navy usage is simply Horizon. For consistency throughout this book, and due to the fact that Japanese instruments are being described, the Japanese term that translates to Artificial Horizon will be used for both services.

Navy Exhaust Gas Analyzer Model 1 manufactured by *Yokogawa Denki Seisakusho*. 80-G-192179.

Exhaust Gas Analyzer

This instrument was popular with Japanese aircraft through World War II, but was nearly phased out of American usage by that time. Its purpose was to increase fuel economy by analyzing the engine exhaust gases and showing the results of this analysis on an indicator scaled in terms of fuel-air ratio (the instrument was often called the Fuel-Air Ratio Gauge). This ranged between 9 to 1 and 15 to 1, the optimum being near the center of the upper scale, where this instrument needle points at 12 to 1. In addition to the fuel-air ratio figures, a parallel scale at the bottom also provided a manifold pressure reading in mm/Hg, same as the Manifold Pressure Gauge. The moving needle on the upper scale would change with the change made by the mixture control, just as the lower scale would be affected by throttle positioning. These two scales were so related that they indicated the optimum mixture ratio for a given manifold pressure. The second scale was used, however, when the correlation data between manifold pressure and fuel-air ratio was available for the particular engine with which the instrument was to be used. As improved carburetors made these corrections automatically, this instrument became less important.

Navy Hydrostatic Fuel Level Gauge manufactured by *Tanaka Keiki Seisakusho*. 80-G-191871.

Hydrostatic Fuel Level Gauge

One of the more complicated instruments to understand was this Hydrostatic Fuel Level Gauge manufactured under license from the American Pioneer Instrument Company. These gauges are readily identifiable by the pump knob at their base. Since it was so widely used in Japanese aircraft it warrants a detailed explanation.

Hydrostatic Fuel Level gauges operate on the principle that the pressure at the bottom of a tank containing fuel is in proportion to the height of the liquid. Hydrostatic means liquid acting as weight, and in this case pressing against air, tending to force the air out of the tank. Referring to the diagram, at the bottom of the tank "T" is a cell "C" that serves as a measuring point for the liquid pressure. The bottom of the cell "C" is connected by small holes to the interior of the main tank, thereby allowing the liquid to enter the cell. The top of the cell is connected by a tube "A" to the inside of the diaphragm "B," which acts as an aneroid, or bellows, for measurement of change.

When the gauge is first installed, the tube "A" will fill with gasoline up to the level of the gasoline in the tank. The pump "D" is therefore operated, and as air is forced into the tube 'A," the gasoline is forced out until the cell "C" is full of air. The pressure that is nec-

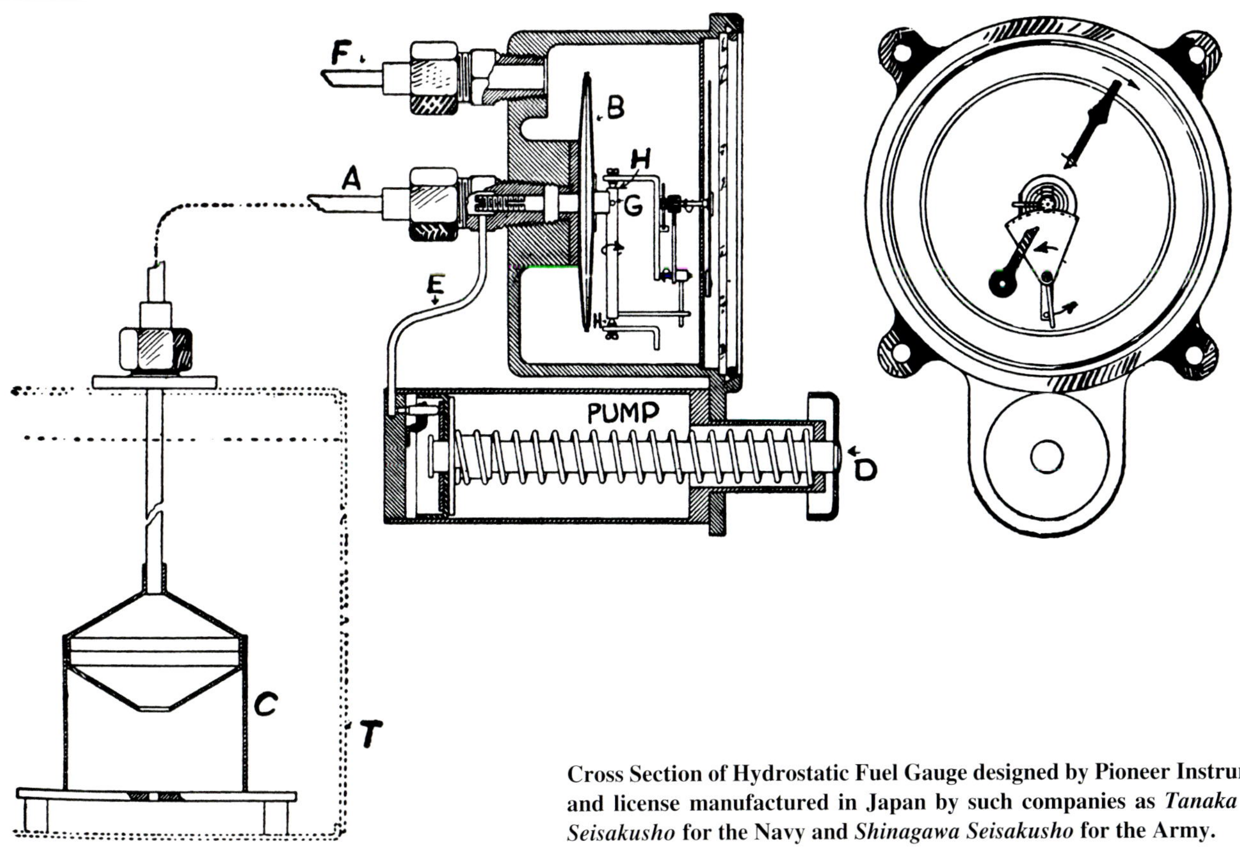

Cross Section of Hydrostatic Fuel Gauge designed by Pioneer Instruments and license manufactured in Japan by such companies as *Tanaka Keiki Seisakusho* for the Navy and *Shinagawa Seisakusho* for the Army.

essary to maintain this air is just enough to keep the gasoline from entering the cell "C" and in proportion to the depth of the liquid. As this pressure is transmitted directly to the inside of the diaphragm through the tube "A," the diaphragm will expand a corresponding amount, and through linkages will rotate an indicating hand to show the amount of fuel in the tank. The gauge always shows the amount of fuel, subject to the errors caused by the slight loss of air. The exact and correct reading may be obtained at any time by a pull of the knob that returns by spring for the pumping cycle.

Manifold Pressure Gauge

This instrument measures the pressure of the fuel-air mixture within the induction manifold that passes into the engine cylinders. The

Manifold Pressure Gauge.

mixture is drawn in by the pumping action of the cylinders. On large engines, the pressure of the fuel-air mixture can be increased for greater power by an impeller, or blower fan connected to the drive shaft at the rear of the engine. The Japanese refer to this blower equipment as a Booster, and the resulting increase in manifold pressure is Boost Pressure; they call the gauge that measures this increase a Boost Gauge. For usage within this book, however, the gauge will be referred to by its American term: Manifold Pressure Gauge.

The Japanese manifold pressure gauge is a familiar looking instrument found in both Army and Navy Japanese aircraft. It reads quite similarly to the British manifold pressure gauge. Both have the same zero point, but the British gauge measures on a scale graduated in pounds per square inch while the Japanese gauge measures on a scale calibrated in millimeters of mercury, abbreviated mm Hg., and needing the addition of a 0 to each number. The scale on an American manifold pressure gauge is calibrated in inches of mercury, abbreviated in.Hg.

With the engine at rest, the needle for the Japanese reading will point close to the 0 mark. This becomes the starting point for measuring from standard sea level atmospheric pressure (absolute pressure), or 760 mm Hg, equivalent to 29.92 inHg found on American manifold pressure gauges. The addition and subtraction in 200 mm Hg increments (less one cipher) has been marked on this gauge from the 0 point. The scale on the following page will help define these readings.

Japanese Gauge	mm Hg	Pressure differences	inch Hg
Red +40	1,160	+400 (mm)	45.67
Red +20	960	+200	37.79
0	760	0	29.92
Black -20	560	-200	22.05
Black -40	360	-400	14.17

(Note: American manifold pressure instruments are graduated in in.Hg relative to the right hand column, but in increments of 10. An average cruise power setting was 20-30, i.e., 2000 rpm and 30 in.Hg.)

When operating Japanese aircraft, pilots were very cautious when "over boosting" into the red area of the Manifold Pressure Gauge. Takeoff and rated power was generally limited to +15 on the red side, depending upon the engine and the aircraft. Pressure beyond that was considered destructive to the engine and only used in critical situations, generally when engaged in combat. Relating this to American engines, they could carry a much higher pressure reading; 52 in Hg being the normal takeoff power for the R-2800 engine in the Douglas B-26 Invader. This would be equivalent to +56 mark (560 mm Hg), which is off the scale of most Japanese gauges.

This instrument panel segment from a Paul shows the red T-handle with inscribed markings that literally reads "Over Boost Pull Handle" that for standardization within this book has been called Emergency Boost Control. Below it is the placard that reads "Over Boost."

Emergency Boost Control

Sometimes referred to as a Manifold Pressure Control, Emergency Power Boost, Supercharger Boost Regulator, Automatic Booster Regulator, and other combinations, depending upon the translator, this unit is used in Japanese aircraft in the high power and performance range. To avoid conflicting terms that all refer to this device, "Emergency Boost Control" will be used. This device should not be confused with the Blower or Supercharger Shift Control, which is worthy of a few words of comment.

Engines of higher horsepower ratings had superchargers that were geared to rotate at higher ratios than the normal blower impellers attached to the crank case at the rear of the engine. These air pumps, if you will, forced more fuel-air mixture into the induction system than was able to be pumped in by the pistons in the engine cylinders, often called *increasing boost pressure*. This greater volume of air mixed with fuel provided more power to the engine. These superchargers, or blowers, often had two speeds, with a lever or knob for manually shifting the clutch mechanism. The higher speed was beneficial at higher altitudes, where the air is thinner and more air was needed.

The Emergency Boost Control, on the other hand, provided a safeguard against over-boost pressure that could easily damage the engine, "blowing a jug" or cylinder being the common term. This activating lever, which was located in the vicinity of the throttle lever—often below or to the rear of it—was normally in the down or back position (This was a pull-out knob on the instrument panel of the Zero, Paul, and other examples.). It functioned through a servo-motor connected to an aneroid near the carburetor. Regardless of how far the throttle was pushed forward, this aneroid would measure the over-pressure limit and prevent the carburetor throttle valve from opening beyond a *rated altitude* limit for the respective engine (For example, the valve setting was plus 25 cm for the Zero). With change in altitude, this aneroid would automatically adjust the carburetor for maintaining that limit. This allowed the pilot to concentrate on air combat where maximum power was the norm, and not have to be concerned about inadvertently over-boosting the engine.

In an emergency situation such as in combat, more power could make the difference between victory or defeat. By moving the Emergency Boost Control lever forward, up, or pulling a handle, the automatically controlled restriction limit was raised (to plus 35 cm for the Zero), and it was then up to the pilot to gamble on the maximum power to apply as viewed on the manifold pressure gauge, and hope that the engine would not fail because of over-boost.

ADI, Water Methanol Injection

Excessive internal heat for large engines when running at very high power was controlled by injecting a 50/50 mixture of methanol and water into the combustion chamber of the cylinders as a form of coolant. Generally this was done automatically, as for example with the Nakajima Ki-84 Frank, this fluid begins to flow at a manifold pressure of +125 mm. Should the water/methanol mixture be exhausted while at high power settings, serious damage to the engine and fluid pump would occur if power is not immediately reduced and the system shut off. It is for this reason that water/methanol pressure gauges, quantity gauges, and low pressure warning lights are installed in most aircraft having this feature. Also known as Anti Detonation Injection, for standardization and brevity for this book, the often used acronym ADI will be used to describe this system and components.

Typical Throttle Control Quadrant when equipped with an Emergency Boost Control that governs excess engine manifold pressure.

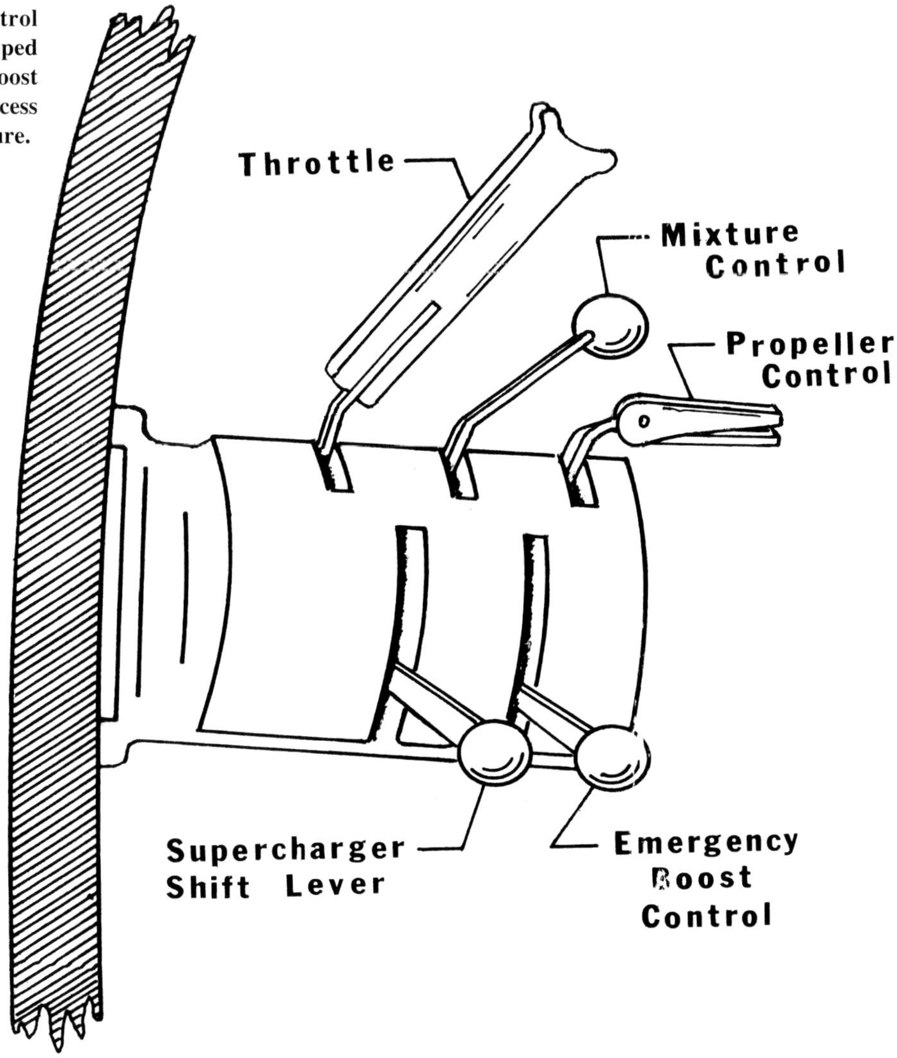

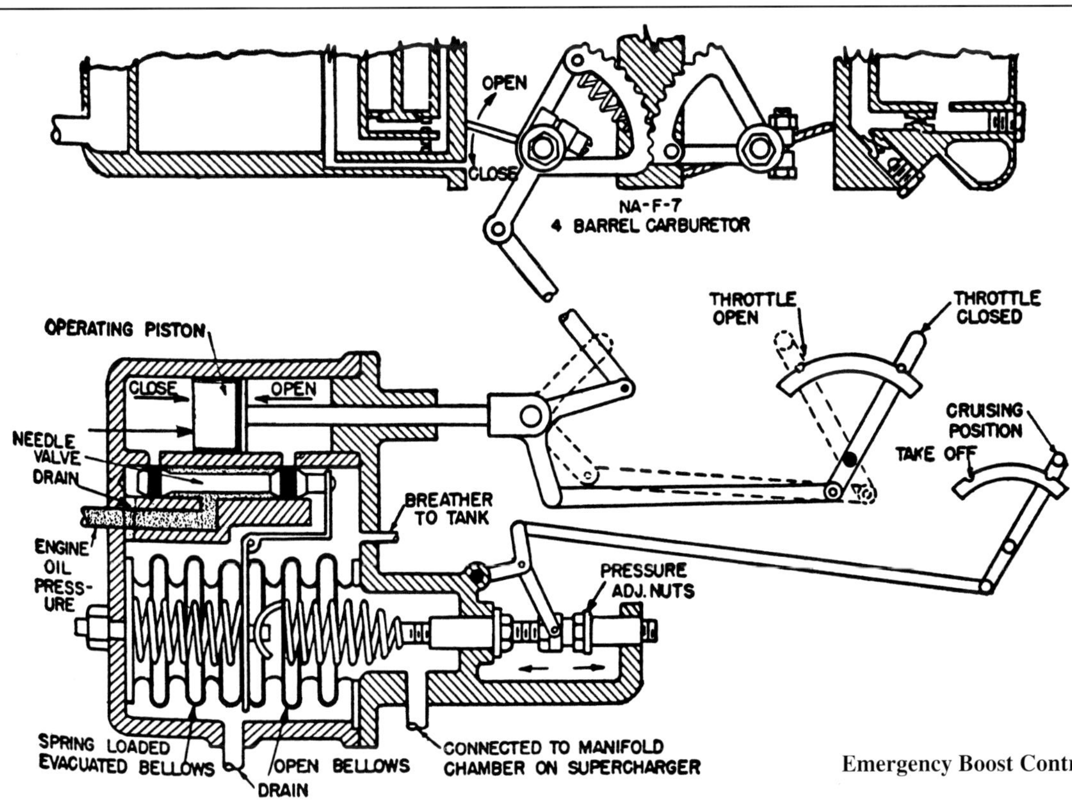

Emergency Boost Control.

Instrument Labeling Placards

A feature that seemed to be more consistent with Japanese military aircraft than any other nation was the labeling of flight and engine instruments on instrument panels. These were very basic names for the respective instruments, not as detailed as giving Model or Type. In normal usage these were thin strips of metal, cardboard, or paper approximately 10 mm high attached to the instrument panel, generally above, yet sometimes below the respective instrument. These varied over the time period of the Pacific War, not only for materials used, but style of characters that were often associated with certain manufacturers for a given time period, or perhaps consistent throughout the war.

During the early war years metal labels were the recognized standard. These were attached with one rivet at each end of the label, then the head carefully painted with the color of the label background if other than silver. Some manufacturers had the characters raised, others indented, in which contrasting paint would reside, but these characters were often machine printed onto the background. Generally, a line of the same color as the characters formed a border around the label. A known exception for having this border was Kawasaki.

As metal became more critical, a lightweight paper card was used in place of metal and continued to be riveted on to the panel. Later these were glued on to the surface. The same style of imprinting the wording was used on the cards as with that of the metal tabs. Late war production used decal-like transfers on some instrument panels. A final form that was observed was stenciling directly to the instrument panel, a seemingly desperate form of reducing one phase of manufacture and saving material.

The observed differences in style used by the two services and their respective aircraft manufacturers are worth noting. This information could lead to identifying an unrecognizable panel containing these labels that are unique to certain manufacturers. The reader must take into account the following are generalizations, and that variations may be noted that do not coincide with what is regarded as the most standard format.

Imperial Japanese Navy

Aichi: Silver background with black characters.
Rounded ends.
Labels that pertained to specific type systems were color coded as follows:
Oil Gauges: Yellow background with white lettering.
Fuel Gauges: Red background with white lettering.
ADI Gauges: Blue or gray backgrounds with white lettering.
Vacuum Gauges: Black background with white lettering.

Kawanishi: Black background with silver characters.
Rounded ends.
Late war production with aircraft such as George used cardboard labels glued in place.

Kugisho: Labels were noticeably inconsistent since most Kugisho aircraft were produced by other companies using their style of labels.

Mitsubishi: Black background with silver characters.
Straight ends.
Labels became decals late in the war, yet retained the same printing format.

Nakajima: Black background with silver characters.
Rounded ends.
Labels became decals late in the war and used the same printing format.
An exception was that of Zero production, for they used the straight end format as Mitsubishi-produced Zeros.

Imperial Japanese Army

Kawasaki: Black background with white characters.
(These labels generally had etched-in characters which were then filled with white paint often being luminescent.)
Borders were lacking.
Rounded ends.

Mitsubishi: Black background with silver or white characters.
Straight ends.

Nakajima: None applied to fighters.
Used in bombers in an undetermined format.

Tachikawa: Black background with white characters.
Rounded ends.
(A known exception was the long-range Ki-77 which used white background with black characters, and having square ends.)

Imperial Japanese Navy

(placards in this section *Courtesy Todd A. Pederson*)

Aichi had a pattern for assigning label color backgrounds to cockpit instruments with regards to systems. This flight instrument label identifying Rate of Climb has silver background with black letters and borders with round ends.

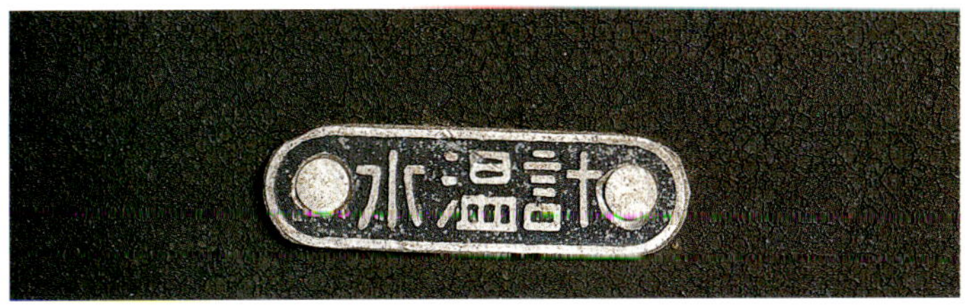

Engine instruments like this Water Temperature Gauge in Aichi aircraft as a rule were the reverse of flight instruments, black background with aluminum or white markings. Exceptions were frequent.

Fuel instrument labels prepared by Aichi could be relied upon for having red backgrounds with white or natural aluminum characters.

These Aichi labels on an early *Seiran* instrument panel provided the standard for the reconstruction of the NASM Seiran. This yellow label identified the Oil Temperature Gauge.

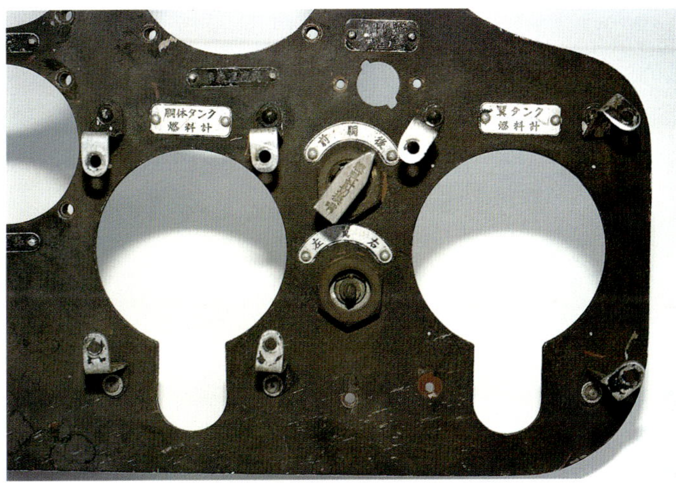

Above: The general practice for Kawanishi (Navy) was the most widely accepted black background instrument labels with aluminum or light gray characters and borders with rounded ends. This one reads Low-Pressure Pressure Gauge. Right: An exception with Kawanishi aircraft is found on the NASM's N1K1 Rex panel, which has the reverse of coloring for this fuel control portion of the instrument panel. Characters were stamped into the metal, then filled with black paint or ink.

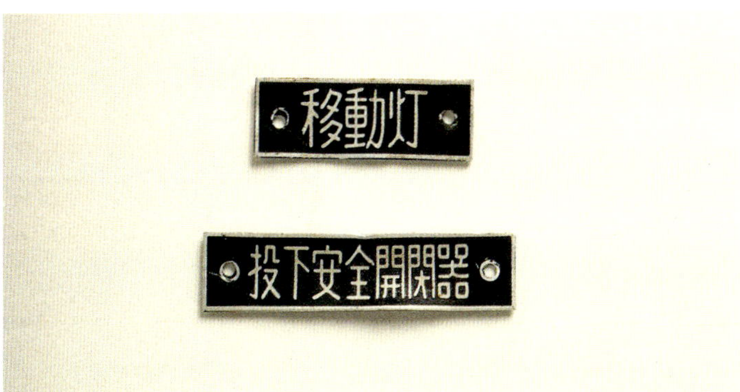

These Mitsubishi (Navy) instrument panel labels were from stock that were never installed. A noticeable mark of Mitsubishi are the straight ends. Black was the normal background, with natural aluminum or painted lettering. Top label is for navigation lights, and lower identified a safety switch to prevent dropping something.

Typical Mitsubishi instrument panel labels are these on an A6M5 Zero panel in the NASM collection. They are straight end with black backgrounds and natural or light gray characters.

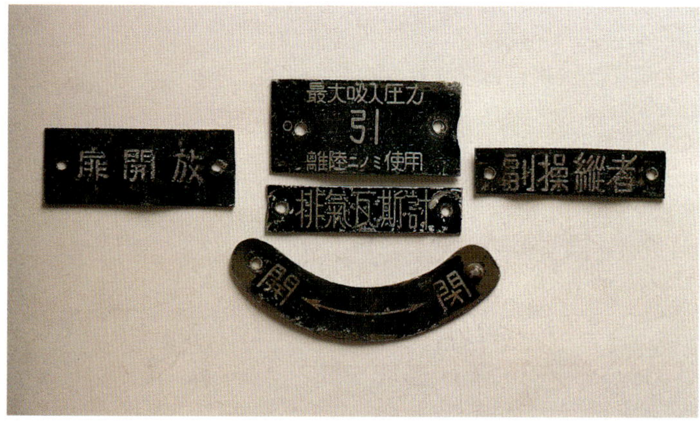

Above: Nakajima (Navy) also used straight end instrument placards for their Navy aircraft. These taken from a J1N1 Irving are crude by comparison. They are without borders, and characters appear to be unpainted aluminum. Left: Late war production by Mitsubishi for the Navy resorted to paper labels with light backgrounds and black characters without borders. The straight ends characteristic of Mitsubishi were retained, yet recognizing that this is a Nakajima built A6M7.

Imperial Japanese Army

(placards in this section *Courtesy Todd A. Pederson*)

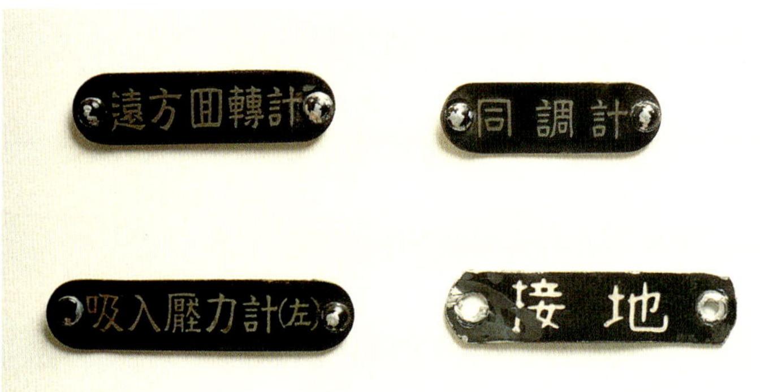

Four examples of instrument panel labels taken from a Ki-48 Lily show two formats in lettering. All have black backgrounds, round ends and are without painted borders. Left to right from top are Remote RPM Gauge, Synchronizer, Left Suction Pressure Gauge, and Contact.

Straight end instrument panel labels found in Mitsubishi (Army) aircraft matched those of Navy production. These placards, said to have come from a Ki-46 Dinah, show inconsistencies in character stroke width, and one has a border while two do not. The two top labels are for Flap Indicator, and at bottom is Forward Edge Tank.

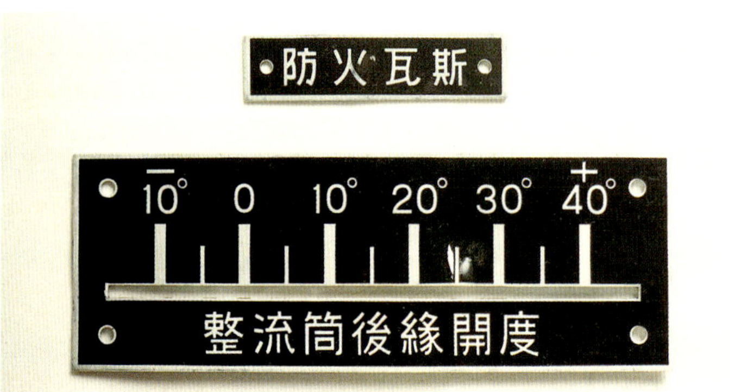

From unused stock are these two labels intended for Mitsubishi Ki-51 Sonia bombers. The style appears to be consistent with Mitsubishi in having black background and straight ends. The top label identified the Fire Protection Gauge, and the lower shows Cowl Flap Opening Position. "O" is trail or neutral pressure position, while -10 is actually "closed" and faired with the fuselage.

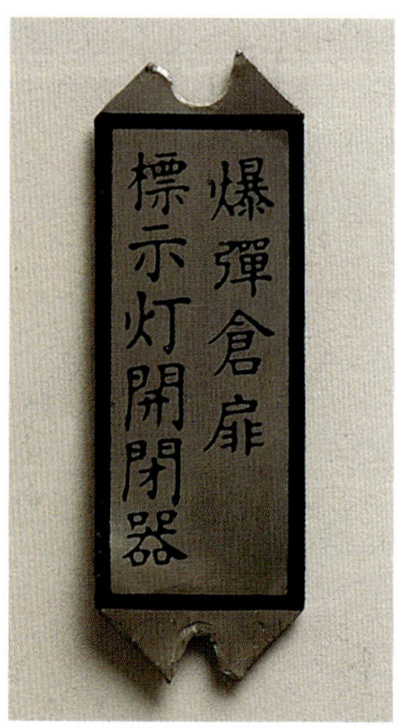

The holder of this instrument label records this having come from a Mitsubishi Ki-21 Sally. This is natural metal with black markings and border. This style is unique from all other known label patterns. Identified as the Bomb Bay Door Open/Close Light Device.

This electrical panel from a Nakajima (Army) Ki-44 Tojo has yet another form of labeling with elongated ends for the rivet. This feature of pointed ends on these labels may be that of the subcontractor and not typically Nakajima. As a general rule, Nakajima Army aircraft instrument panels did not contain labels.

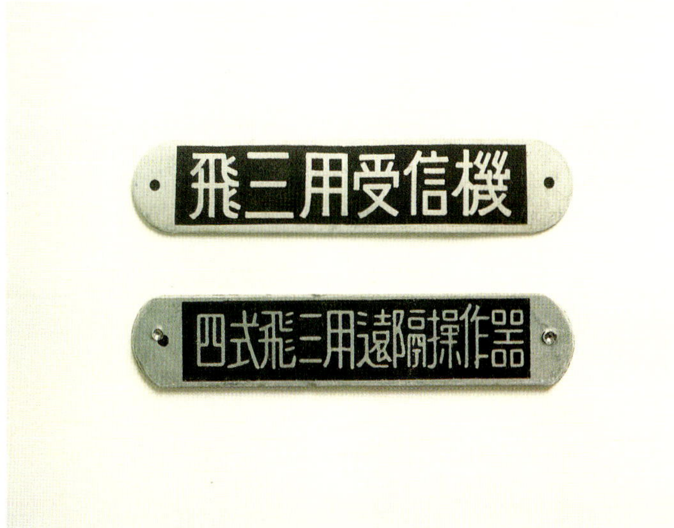

These two labels from stock identify the receiver for the *Hi* 3 radio, and a Type 4 Hi 3 Remote Actuation Device; both are shortened identifiers that were intended for a Nakajima Ki-84 Frank. These were unusual, in that they have straight end black backgrounds, yet rounded ends.

Another electrical panel, this one from a Nakajima Ki-84 Frank, has ivory color plastic labels with black lettering. This is an indication of conserving metal in this late war production aircraft.

Instrumentation

This collection of instrument photos on file in the National Archives constitutes a cross section of typical Japanese World War II aircraft instruments.

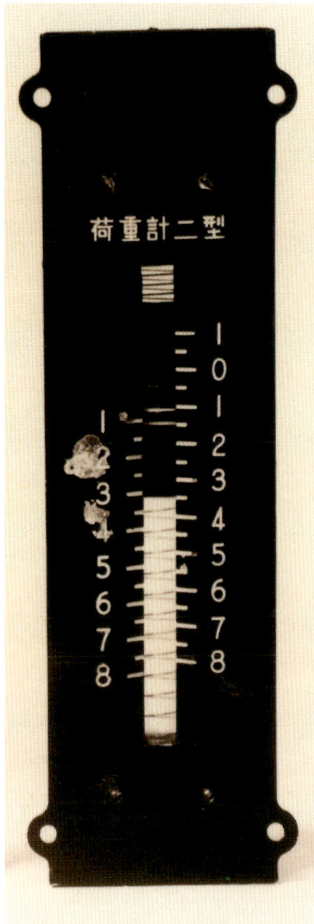

ACCELEROMETERS were a relatively new instrument at this time of aviation. This Navy Type 2 found in Frances was rare for a bomber needing an instrument to read positive and negative G (gravity) forces exerted on an airframe. But Frances was an attack bomber, subject to dive recovery forces recorded by this instrument. *80-G-192539.*

ACCELEROMETER. Close-up view of an Army Type 1 Accelerometer of unknown manufacture. Appearing as one needle here, since both are at the 1 'g' reading while at rest, one will record negative 'g' force, while the second needle records positive 'g' force. *80-G-192195.*

ALTIMETER This Navy Simple Altimeter Model 3 was standard with many Japanese Navy aircraft. Calibrated in thousands of meters, they were graduated to 8,000 m (26,250 ft). Manufactured by *Tanaka Keiki Seisakusho* and Katsura Kenkyusho. *80-G-191874.*

ALTIMETER. This is a Navy Sensitive Altimeter Model 2 used on most operational Naval aircraft. These gauges, manufactured by *Tanaka Keiki Seisakusho* and Katsura Kenkyusho, had what on American instruments was the Kollsman window for setting the local barometric pressure for more accurate altitude readings. *80-G-192335.*

ALTIMETER. Intended for low performance aircraft, this Army Type 95 Model 2 Altimeter reads 5,000 m (16,405 ft) with one sweep of the hand, and up to 10,000 m on the second full turn. Built by *Yanagi Seisakusho. 80-G-191882.*

ALTIMETER: This is the Army Type 97 Altimeter built by *Tanaka Keiki Seisakusho*. The thumb knob adjusts for changes in barometric pressure. This instrument has two hands, the larger hand reads 1,000 m (3,281 ft) with one revolution, and 10,000 m (32,810 ft) with one revolution of the small needle. *80-G-191832.*

AIRSPEED INDICATOR. This is a Navy Airspeed Indicator Model 3 manufactured by *Tanaka Keiki Seisakusho* used in many of the heavy type Navy aircraft. *80-G-191915.*

AIRSPEED INDICATOR. An improvement is this Navy Airspeed Indicator Model 3 *Kai* 1 manufactured by Tanaka Keiki Seisakusho. This type was to be used in Kikka and other late war aircraft. *80-G-191860*

AIRSPEED INDICATOR. A later version of the Navy Airspeed Indicator Model 3 *Kai* 1 had a higher range of air speed than its predecessor with the same designation. *80-G-191834*

AIRSPEED INDICATOR. This is a detail view of the Navy Airspeed Indicator Model 6 manufactured by *Tanaka Keiki Seisakusho*. Its 600 kt (325 kt) was a much higher reading than other standard airspeed indicators. *80-G-192544*

AIRSPEED INDICATOR. Army Type 98 Airspeed Indicator built by *Tanaka Keiki Seisakusho*. Maximum reading shown is 700 km/hr (380 Kt) on the inner scale to be read on the second sweep of the needle. *80-G-191843*

AIRSPEED INDICATOR. This dismounted unit is an Airspeed Line Clearing Pump manufactured by *Tanaka Keiki Seisakusho*. A plunger type device of this type is actuated to clear condensation from the pitot static line. These are normally mounted on the instrument panel. *80-G-191881*

ARTIFICAL HORIZON. This Navy Model 1 Artificial Horizon was manufactured by *Tokyo Koku Keiki* under license from the American firm Sperry. An identifying mark of the Sperry design is the high wing configuration on the miniature airplane with an extended landing gear and a blue background on the sky area. *80-G-191872*

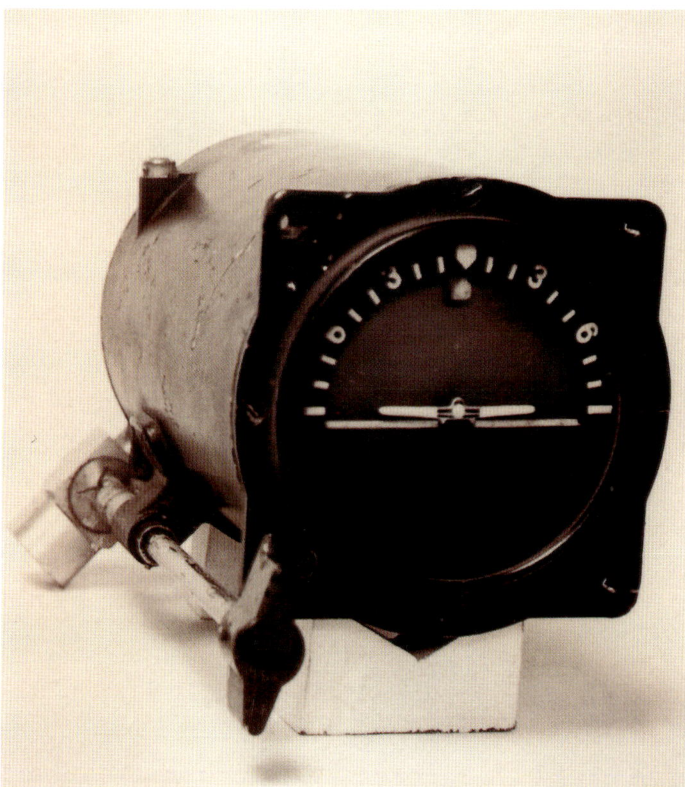

ARTIFICAL HORIZON. Several types of artificial horizon gyro instruments were used by the IJN during the Pacific War. This Artificial Horizon Model 1 manufactured by *Tokyo Koku Keiki* was used in Betty and others of that time period and class. *80-G-191839*

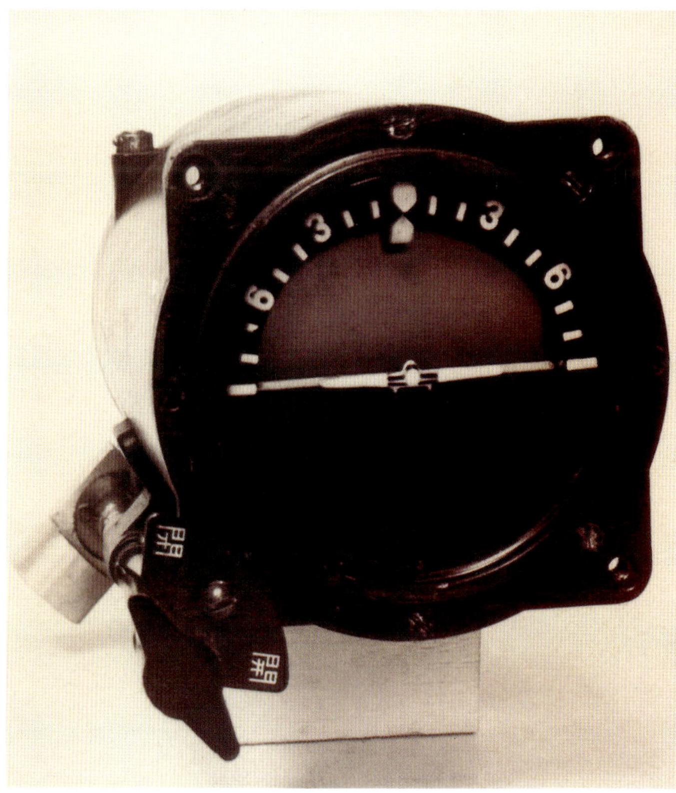

ARTIFICIAL HORIZON. An improved model was this Navy Artificial Horizon Model 2 by *Tokyo Koku Keiki*. The caging switch for these gyro instruments is at the lower left. *80-G-192297*

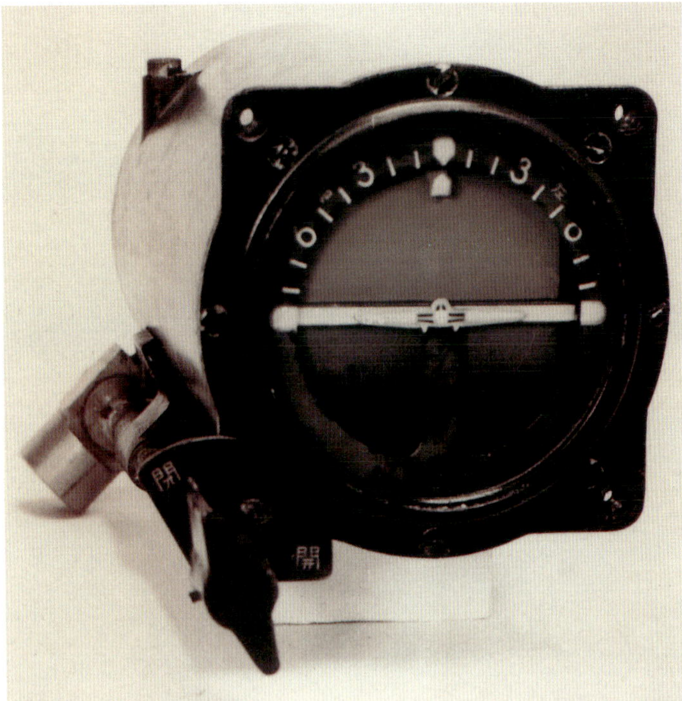

ARTIFICIAL HORIZON. Another version of the Navy Artificial Horizon Model 2 had a variation in the horizon bar and the wings-level markers. ***80-G-191858***

AUTOPILOT. This is the gyro compass portion of the Navy Type 1 Autopilot. The bottom circular card is set by turning the lower center knob to the desired heading of flight. The upper rotating card is the gyro compass heading that is to align the aircraft with the heading set in the direction card at the bottom. ***80-G-192489***

AUTOPILOT. This combination of the directional gyro and artificial horizon make this the Navy Model 1, Type 1 Autopilot. These components were manufactured by ***Tokyo Koku Keiki. 80-G191903***

AUTOPILOT. This is the standard Army Type 95 Autopilot, also used by the Navy as Model 0. Built by *Tokyo Koku Keiki*, this was originally a license built copy of the Sperry Automatic Gyropilot predominantly used on most WWII American multi-engine aircraft of the late 1930s and 1940s. This was hydraulically operated and very reliable. *80-G-191900*

Right: AUTOPILOT. This rack supported the two major autopilot controls mounted within the instrument panel. At its base are the three sensitivity controls for (L to R) rudder, aileron, and elevator. The gauge below indicates hydraulic pressure for the operation of the system. A separate left-right valve turned the autopilot on and off. *80-G-192181* **Below left: DIRECTIONAL GYRO.** A companion gyro to the horizon is this Navy Directional Gyro Model 1 Manufactured by *Tokyo Koku Keiki* from a license granted by Sperry. The caging knob at the bottom is also used to align the compass card with a magnetic compass. Gyro precession required this alignment to be done about every 15 to 20 minutes. This early version had large numbers on the compass card and wide scale marks. *80-G-191848* **Below right: DIRECTIONAL GYRO.** This Army Directional Gyro built by *Tokyo Koku Keiki* was not identified by any Type or Model markings..Being relatively standard, its caging knob in the center also aligned the compass card manually with the magnetic compass. *80-G-192870*

COMPASS. A standard Navy compass found in many aircraft was this Type 92 Compass Model 2. This instrument was manufactured by *Yokogawa Kenki Seisakusho* beginning in 1938. The outer heading ring could be manually rotated to set the desired heading course at the top for a constant reference. *80-G-191912*

COMPASS. This is a Navy Type 94 Induction Compass manufactured by *Yokagawa Denki*. This compass used a power-driven coil revolving in the earth's magnetic field as a sensing element. This was formerly known as an earth-indicator compass. *80-G-192568*

COMPASS. This is the more conventional Navy Type 0, Model 1 compass manufactured by *Tokyo Koku Keiki*. At the bottom of the case was a magnetic deviation adjustment scale. *80-G-192169*

COMPASS. An improved version of the Type 0 was this Navy Type 1 Magnetic Compass Model 2. Normally one was placed in each flight crew position in larger aircraft. These were manufactured by *Tokyo Koku Keiki*. *80-G-191930*

COMPASS. This Navy Navigator's Compass Model 1 looks cumbersome compared to others. It is a vertical reading compass, having a series of mirrors so that it can be read when looking horizontally. In aircraft such as the Nakajima C6N Myrt, these were placed in a recess of the pilot's instrument panel for his use. *Tokyo Koku Keiki* was the manufacturer. *80-G-191931*

COMPASS. This Navy projection type compass is a type known to have been used in early bombers, such as the Mitsubishi G3M Nell. These floor mounted instruments were identified as being manufactured by C. Plath in Germany. *80-G-192570*

COMPASS. This Army Type 98 *Ko* Magnetic Compass was a commonly used magnetic compass throughout the war. These were manufactured by *Tokyo Koku Keiki*. An adjustment for compass deviation is at the top of the case. *80-G-191932*

COMPASS. This Army Type 98 *Otsu* Magnetic Compass is very similar to the Navy's Type 0 Model 1, with subtle differences. These were also manufactured by *Tokyo Koku Keiki* and widely used. *80-G192170*

COMPASS. This vertical reading Navy Type 2 Navigator's Compass Model 2 was commonly found at navigator's stations in bombers, but also floor mounted forward of the control stick in some fighters and single engine bombers. *80-G-192572*

RATE OF CLIMB. This basic flight instrument is a Navy Rate of Climb Model 1 manufactured by *Tokyo Koku Keiki*. Vertical speed is measured by meters per minute, the 500 meter rate being the most standard under instrument flight conditions. *80-G-191877*

RATE OF CLIMB. Similar to the Navy model is this Army Type 97 Rate of Climb, also manufactured by *Tokyo Koku Keiki*. This early type has more defined calibrations better suited for fighter aircraft. This type was found in the Oscar flight tested at Eagle Farms in Brisbane.

TURN AND BANK. Another standard flight instrument was this Navy Turn and Bank Indicator Model 2 built by *Tokyo Koku Keiki*. Used also in visual flight conditions, it measured the rate of turn for the aircraft. To coordinate the turn so as not to slip or skid, rudder pedal pressure centered the black ball in the tube filled with fluid for dampening. *80-G-191838*

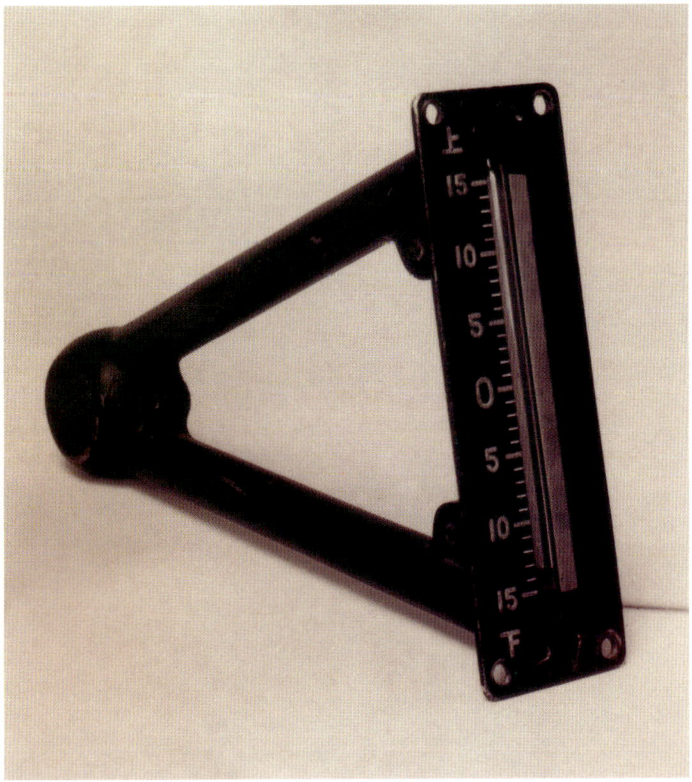

INCLINOMETER Model 2 was used for measuring the angle of dive and was found in all the IJN carrier attack aircraft. These were simple instruments with no moving parts except for the liquid and air in the triangular-connected tube This measured the dive angle, as well as a climb angle, much like a carpenter's level. *Tokyo Koku Keiki* manufactured these instruments. *80-G-191931*

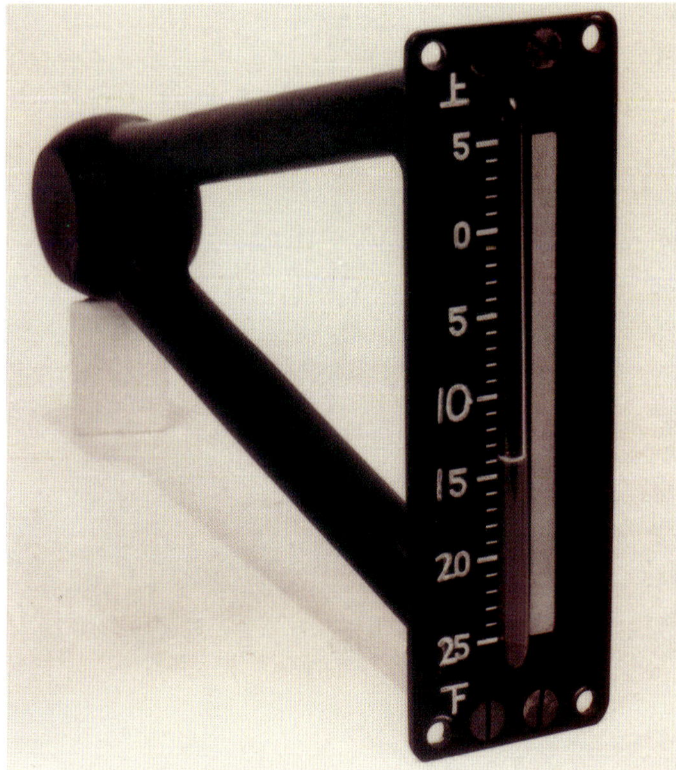

INCLINOMETER. This Inclinometer Model 2 used in *Ohkas* was manufactured by *Tokyo Koku Keiki* Seisakusho. It differs from the standard by having a modified scale, placing the zero mark near the top for making steeper dive angles. *80-G-192676*

TACHOMETER. Cable driven tachometers such as this were of the type used in lightly powered aircraft, such as the Kugisho K5Y Willow. This is a Mk.1 Chronometric Tachometer Model 1 manufactured by *Tanaka Keiki Seisakusho*. *80-G-192178*

TACHOMETER. One of several tachometers used by the Japanese Navy was this Mark 1 Chronometric Tachometer Model 1 found in the Aichi E13A Jake. Manufactured by *Tanaka Keiki Seisakusho*. These were mechanically cable-driven. *80-G-192304*

TACHOMETER. Engine speed could be measured accurately by this Navy Electro-Tachometer Model 1, manufactured by *Yokogawa Denki Seisakusho*. This type tachometer was widely used on the more advanced aircraft. *80-G-191907*

TACHOMETER. A departure from the more popular western nations that used round-face tachometers with two needles when used for twin engine aircraft, Japanese designs followed the vertical reading approach. This is the Dual Electronic Tachometer Model 2 Kai (Improvement) 2 used in Bettys, Frances, and others. These were made by *Yokogawa Denki Seisakusho*, as well as Dai Nippon Kaigun Kokusho Heikibu. *80-G-192678*

TACHOMETER. There were several models of Dual Electronic Tachometers designated Model 2 Kai 2. Their external differences appeared to be in the layout of graduation marks. Instrument panel photos of Emily show this to be the design used for that aircraft. Manufactured by *Yokogawa Denki Seisakusho*. *80-G-192573*

TACHOMETER. This is an Army Type 14 Chronometric Tachometer Model 2 manufactured by *Tanaka Keiki Seisakusho*. These gauges were used on early war period fighters, such as Oscar 1 and lesser powered aircraft. *80-G-191880*

TACHOMETER. Heavier powered aircraft used instruments such as this Army Electro Tachometer Type 100 for reliability. This model was manufactured by *Yokogawa Denki Seisakusho*. *80-G-191873*

TACHOMETER. One of several vertical reading tachometers with dual scales is this Army Type 98. A pointer on each side recorded respective engine speeds for twin engine aircraft. These were manufactured by *Yokogawa Denki Seisakusho*, as well as *Mitsubishi Denki*. *80-G-191911*

MANIFOLD PRESSURE. This Navy Manifold Pressure Gauge Model 2 was obviously designed for lower powered engines, such as the 875 hp Zuisei 13 engine that powered the F1M2 Pete. Its high manifold pressure range was limited because of its single speed, low gear-ratio blower. This gauge was manufactured by *Tanaka Keiki Seisakusho*. *80-G 191835*

MANIFOLD PRESSURE. This is a Navy Mk.1 Manifold Pressure Gauge Model 3 produced by *Tanaka Keiki Seisakusho*. This model was standard for the higher powered aircraft, such as the J2M3 Raiden. *80-G-191909*

MANIFOLD PRESSURE. Duel reading gauges were new by only a few years in Japanese aircraft. Twin-engine Navy aircraft used one of these dual Navy Type 4 Manifold Pressure Gauge Mk.2. These were produced by *Tanaka Keiki Seisakusho*. *80-G-191865*

MANIFOLD PRESSURE. A standard Army instrument was this Type 94 Manifold Pressure Gauge that was used in aircraft such as the Ida, front and back seats. *Tanaka Keiki Seisakusho* was the manufacturer of this instrument. *80-G-191875*

MANIFOLD PRESSURE. A standard instrument for Army manifold Pressure Gauges was this Type 98 built by *Tanaka Keiki Seisakusho*. The "0" reading is close to standard sea level barometric pressure, and the rest of the instrument is calibrated Kg/cm^2. *80-G-191866*

CYLINDER HEAD TEMPERATURE. This Army Cylinder Head Temperature Gauge Model 1 was a fairly common instrument for this vintage aircraft. These were manufactured by *Yokogawa Denki Seisakusho*, *Mitsubishi Denki*, and *Fuji Koku Keiki*. *80-G-192196*

CYLINDER HEAD TEMPERATURE. This dual-reading Cylinder Head Temperature Gauge was designed to operate with twin-row cylinder engines. These utilized a wafer switch at the base of the instrument to select a reading from one of the front row cylinders or one from the rear row of cylinders. These were manufactured by *Fuji Koku Keiki*. *80-G-192302*

CYLINDER HEAD TEMPERATURE. As airplanes grew in number of engines, instruments had to consolidate in size. Two of these Navy Exp. Type 4 Mk 2 Dual Cylinder Head Temperature Gauges handled the four engines that powered the Nakajima G8N1 Rita. These were manufactured by *Fuji Koku Keiki*. *80-G-192341*

ADI PRESSURE. To prevent detonation at high engine power settings, a water-methanol injection into the cylinders for cooling was devised. It was critical that this limited supply of liquid not become exhausted during these high settings. This Navy Mk.2 ADI Pressure Gauge Model 4 was used to monitor this operation. *80-G-192294*

ADI PRESSURE. This Navy Dual ADI Pressure Gauge Mk.2 Model 4 could monitor two engines. One of these was found in the twin-jet powered *Kikka*, but its purpose in that type aircraft is unknown. *80-G-192292*

COOLANT TEMPERATURE. There were several designs for faces of coolant temperature instruments. Among the few liquid cooled engine powered aircraft that would need them were the Tony, Judy, and *Seiran*. *80-G-192542*

EGT. This is a Navy Dual Exhaust Gas Temperature Gauge, Type 4, Model 2 Mk.2 manufactured by *Fuji Koku Keiki*. Aircraft known to have used this type instrument were Frances, Irving, Lorna, and Rita. *80-G-191901*

EGT. This was the standard Navy Exhaust Gas Temperature Gauge found on most operational aircraft. It was used in leaning the fuel mixture for the engine to give best performance and duration. Several manufacturers made this instrument, including *Mitsubishi Denki*, *Yokogawa Denki Seisakusho*, and *Fuji Koku Keiki*. *80-G-192339*

EGT. The Army used a different format than the Navy, although these instruments were made by the same manufacturer, *Fuji Koku Keiki*. Ths is the Army Type 2 Exhaust Gas Temperature Gauge. *80-G-191833*

EXHAUST GAS ANALYZER: Army Type 2 Exhaust Gas Analyzer built by *Yokogawa Denki Seisakusho* is shown here. Its purpose was for obtaining the most efficient fuel consumption. The top pointer showed the fuel-to-air-ratio, which was adjusted by the mixture control as compared to the manifold pressure on the lower pointer, which was controlled by the throttle. *80-G-191842*

CARBURETOR AIR TEMP. This gauge measured the temperature for determining the proximity of which ice may form in the carburetor. These were manufactured by *Fuji Koku Keiki*, *Yokogawa Denki Seisakusho*, and *Mitsubishi Denki*. *80-G-192569*

FREE AIR TEMPERATURE. This Free Air Temperature Gauge was a common item in the heavier Japanese Navy aircraft. This gauge was also installed in other crew member positions besides the pilot. Most were manufactured by *Yokogawa Denki*, *Seisakusho*, and *Mitsubishi Denki*. *80-G-191859*

CARBURETOR AIR TEMPERATURE. This Navy Air Temperature Gauge Model 4 (Dual) would be standard on most late war Navy aircraft. This measures the engine inlet air temperature before and after carburation. These were manufactured by *Fuji Koku* Keiki. *80-G-192203*

CARBURETOR AIR TEMPERATURE. This dual carburetor Air Temperature Gauge gives readings of the air passing through the carburetor, often necessary to know in moist conditions in order to watch for carburetor icing and prevent it by adding carburetor heat. This two position instrument for two stages of air passage was manufactured by *Fuji Koku Keiki*. *80-G-191846*

OIL TEMPERATURE. This is the standard Navy Mk.1 Oil Temperature Gauge Model 5 found in most Navy aircraft. These were manufactured by *Shiki Keiki*. *80-G-191868*

OIL TEMPERATURE. Another dual instrument so necessary for the confines of twin-engine single-pilot aircraft like Irving. This is an unnumbered Navy Oil Temperature Gauge, manufacturer unknown. *80-G-192332*

OIL TEMPERATURE. A common engine gauge is this Army Type 98 Electric Oil Temperature Gauge. These were manufactured by *Fuji Koku Keiki*. *80-G-191869*

OIL TEMPERATURE. This is a 60 mm diameter dual Oil Temperature Gauge found on most Army multi-engine aircraft. These were manufactured by *Shinagawa Seisakusho*. *80-G-192333*

OIL PRESSURE. This Navy Oil Pressure Gauge Model 1 Mk.1, manufactured by *Tanaka Keiki Seisakusho*, was common among most Navy aircraft. *80-G-192546*

PRESSURE. Very similar to its predecessor, this Navy Pressure Gauge Model 2 does not specify being exclusively for oil pressure. Its indices give the actual figure without having to add the cipher. These were manufactured by *Tanaka Kieki* Seisakusho. *80-G-192334*

OIL PRESSURE. This is an example of one of four Navy Oil Pressure Gauge Model 4 seen on the instrument panel of Emily. These were manufactured by *Tanaka Keiki Seisakusho*. *80-G-192556*

OIL PRESSURE. A standard instrument for the Army was this Army Type 94 Oil Pressure Gauge found in most single engine higher powered aircraft. These were manufactured by *Shinagawa Seiskusho*. *80-G-192192*

OIL PRESSURE. The Army Type 2 Oil Pressure Gauge (Dual) was designed to save space in twin engine aircraft. Manufactured by *Shinagawa Seisakusho,* this type can be seen in the Nick fighter. *80-G-192190*

OIL PRESSURE. Another dual Army oil pressure gauge is this one manufactured by *Shinagawa Seisakusho*. Note the change in character for "oil." *80-G-192191*

ENGINE GROUP. Pressure Gauges of this type were a standard Navy instrument for many of its aircraft. This Mk.2 Model 1 records oil pressure at left and fuel pressure at right. As was industrial practice in Japan, readings were measured in kg per square cm. These were manufactured by *Tanaka Keiki Seisakusho*. *80-G-191840*

ENGINE GROUP. This Navy Engine Gauge Unit 15-Shi gives oil temperature at top in C x 10, oil pressure at left, and fuel pressure at right graduated in kg per square cm. The 15-*Shi* means an acceptance of this type in Showa 15 (1940). *80-G-192204*

HYDRAULIC PRESSURE. This was the standard Army Hydraulic Pressure Gauge in general use in Army aircraft. It can best be seen on the Oscar 2 instrument panel layout as Item 16. These were made by *Shinagawa Seisakusho*. *80-G-191837*

HYDRAULIC PRESSURE. This Navy Type dual hydraulic pressure gauge carried no additional identification. Depending on system pressure, it could have read pressure from each of the engine driven pumps or perhaps break pressure. *80-G-192554*

HYDRAULIC PRESSURE. Another dual reading instrument, but one not specifically marked as to function. Marked as Navy Model 1 by *Tanaka Keiki Seisakusho*, it is most likely hydraulic pressure and vacuum Gauge. *Courtesy of Todd Pederson.*

BRAKE PRESSURE. This is one of several early dual-type reading instruments for multi-engine Japanese aircraft. Normally two of the same instruments were used. Shown here is a Dual Hydraulic Brake Pressure Gauge manufactured by *Shinagawa Seisakusho* of the type likely to have been used for the Nakajima Ki-67 Peggy. *80-G-192189*

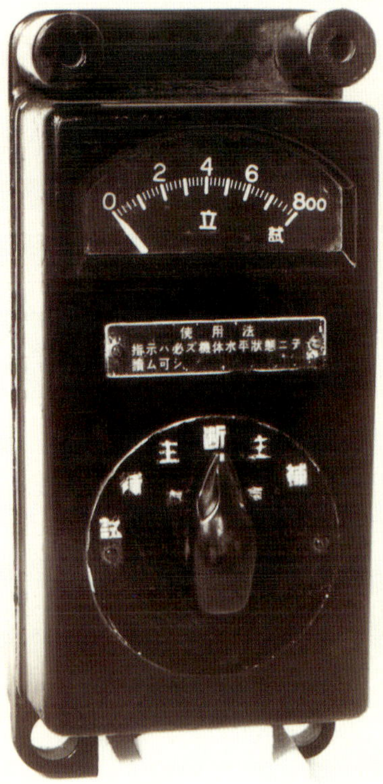

FUEL QUANTITY SWITCH. Navy Type Liquidometer Fuel Gauge manufactured by *Fuji Koku Keiki*. This gauge imitates the American liquidometer in the Douglas DC-2, of which Japan had production rights. The four center-section fuel tanks were nearly the same capacity (200 U.S. Gal), therefore this one quantity scale for 800 liters was suited to all four tanks. *80-G-192328*

FUEL QUANTITY SWITCH. One fuel gauge could measure any of several fuel tanks by having a Fuel Tank Reading Selector Switch as shown here. This is a six position switch like one suitable for the Nakajima Ki-49 Helen to selectively monitor individual tanks. This one was made by *Fuji Koku Keiki*. *80-G-192869*

FUEL QUANTITY. As modern as jet aircraft were at the time the war ended, it is interesting to note the earlier hydro-static fuel quantity gauges of the Pioneer licensed manufacturer variety were used in the jet powered *Kikka*. This one was a product of *Tanaka Keiki Seisakusho*. Quantity numbers are present but not visible here. *80-G-191867*

FUEL QUANTITY. This Navy Electric Fuel Gauge is calibrated to read more than one tank capacity. Seemingly, this aircraft had five fuel tanks for which their filled capacity are marked by their circled tank number. *80-G-192337*

FUEL QUANTITY. A more standard form of instrument is this Army Type 98 Electric Fuel Gauge built by *Fuji Koku Keiki*. *80-G-192198*

FUEL QUANTITY. Of the electric hydrometer type fuel quantity gauges was this Army Type 97, which reads in 100s of liters. These were manufactured by *Mitsubishi Denki*. These are usually accompanied by a selector switch so that this one gauge can read tank quantities selectively. *80-G-191836*

FUEL QUANTITY. This earlier type gauge is an Army Type 100 Fuel Gauge manufactured by *Shinagawa Seisakusho*. It is a hydro static type whose function is described earlier in this chapter. *80-G-191870*

FUEL PRESSURE. This Navy Fuel Pressure Gauge Model 2 was built by *Tanaka Keiki Seisakusho*. Two of these would be found on the Tabby panel. *80-G-192200*

FUEL PRESSURE. This is an Army Type 98 Fuel Pressure Gauge in general use at the start of the war. These were manufactured by *Shimaza Seisakusho*. Of note are the decimal points at the top of the numbers. *80-G-191841*

FUEL PRESSURE. Army Dual Fuel Pressure Gauge of unknown manufacturer. This instrument is typical for most Army multi engine aircraft. One can be seen on the Sally panel. *80-G-192549*

FUEL PRESSURE.. Lower pressure readings are on this Army Pressure Gauge manufactured by *Shinagawa Seisakusho*. It contains no other identification. An instrument of this type was found in Tony. *80-G-192199*

VACUUM. At the time that instrument flying was practiced more in the late 1930s, air driven gyro instruments were being driven by engine driven vacuum pumps, as opposed to externally mounted venture tubes. That source of power was measured by this type of Vacuum Pressure Gauge Model 1 built by *Tokyo Keiki Seisakusho*. *80-G-192336*

VACUUM. This Navy Vacuum Gauge Model 1 is different from the preceding, but has the same nomenclature. This was manufactured by *Tokyo Keiki Seisakusho* and was installed in a Val. *80-G-192293*

OXYGEN. This Navy Oxygen Flow Regulator Model 2 was manufactured by *Tanaka Keiki Seisakusho*. This was standard for many Japanese Naval aircraft. *80-G-191855*

SYNCHRONIZER. Propellers out of synchronization produce an annoying beat for the crew members. Minute adjustment by sound can match their RPM without the need for an instrument. However, when passing the neutral point it is difficult to know if the process has gone past neutral or reversed at mid-point. An instrument such as this Army Type 100 Propeller Synchronizer Gauge can show the direction and the point of synchronization so as to make the correct adjustment with the throttles. These were manufactured by *Yokogawa Denki*, as well as *Mitsubishi Denki*. *80-G-191847*

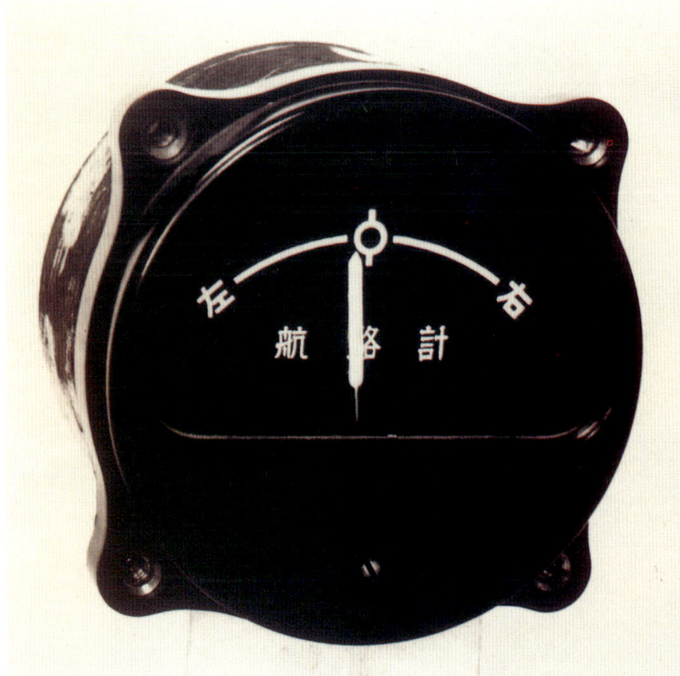

RADIO COMPASS indicators like this operated by signal strength obtained from the loop antenna. The center needle gave direction left or right to the transmitting station. The left needle indicated intensity of the signal so as to determine if going to or from the station. Manufactured for the Navy by *Yokogawa Denki Seisakusho*. When installed, this unit was generally near the center of the instrument panel. *80-G-192665*

HOMING DIRECTION. Another type of instrument for navigation is this Navy Type 1 Ku Model 3 Homing Direction Finder. This system also homes in on radio signal strength, and the needle deflects left or right depending on the direction to turn to center the needle. Manufacturer is unknown. *80-G-192553*

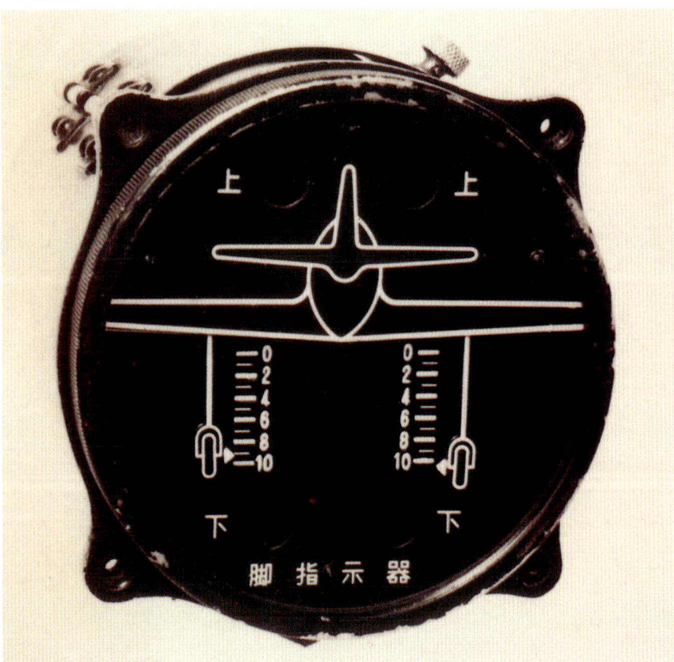

Above: LANDING GEAR POSITION. There were many designs used for showing the landing gear position to the pilot. This was the type most likely used on Helen, one that shows intermediate positions between full up and down-and-locked. This instrument was manufactured by *Fuji Koku Keiki*. *80-G-192330* Right: FLAP INDICATOR. This is a Landing Flap Position Indicator that shows maximum deflection for this unidentified aircraft of 37 degrees. Most likely used on an Aichi (Navy) aircraft, since the instrument is by *Aichi Kokuki*. *80-G-191845*

FLAP INDICATOR. Each aircraft type seemed to have its own position indicators. This Landing Flap Position Indicator built by *Nippon Seimitsu Denki* is from an unknown aircraft. *80-G-192303*

FLAP INDICATOR. Another form of flap deflection indicator was this rotating vernier scale to indicate deflection up to 60 degrees. Type of Navy aircraft is unknown. *80-G-192548*

DIVE FLAP. This Dive Flap Position Indicator is unique to the Frances. It was manufactured by *Nippon Seimitsu Denki*. In the cockpit it was located on the auxiliary instrument panel at the left of the pilot. *80-G-192300*

3

Radio Equipment

With the assistance of
Z.I. "John" Szewezyk of ARTA* and Todd A. Pederson

Japanese radio material was regarded by Allied forces during the Pacific War as varying from poor to excellent. Early captured equipment had revealed outdated designs and construction that appeared to be several years behind American technology. From the mid-point of the war marked improvement in Japanese radio design was noted. Some equipment manufactured in late 1943, such as the Navy Type 3*Ku* Mk.1 Wireless Telephone, had the latest features of electronic design and showed excellence in construction. Japanese frequence crystals were of good quality. Throat microphones manufactured for the Japanese Navy were evaluated as superior to USAAF types.

The reliance upon the use of radios was considerably different between the U.S. and Japan. Before and during this wartime period, Americans valued the use of voice radios for local traffic control at airports, *en route* navigation position reports, and military voice communication between airplanes. Because of this, aircraft radio equipment had to have multiple frequencies that could be changed during flight. Until multichannel VHF radios with eight push-button pre-selected channels became available some time in 1943, frequency changes for U.S. aircraft radios were made with what was known as the "coffee grinder," a small hand cranked frequency tuner for the transmitter and one for the receiver. Radio tuning was a tedious task for pilots in single place aircraft.

The Japanese military, on the other hand, stressed radio silence, and thus with very little voice traffic, few, if even more than one radio channel was needed. Accepted reasons for making radio transmissions were to alert one's comrades of enemy attack, to give an order for an attack, or to report the sighting of major enemy activity. For those needs, a single Receiver/Transmitter channel, crystal controlled radio, pre-tuned before flight, relieved the pilot of this tuning chore and distraction from piloting the airplane. This did, however, restrict the Japanese from taking greater advantage of voice communication. Instead, the Japanese military preferred radios that were small, simple, and lightweight, and the industry gave them just that.

The development of radar during World War II changed much of this Japanese concept. Radio silence would no longer be part of the element of surprise, since radar would detect the approaching enemy aircraft. Radio jamming by the Americans to the single operating radio frequency of the Japanese often made these radios useless, since there were no alternate frequencies to tune to. These shortcomings were recognized by European nations prior to the war, but these threats were discounted by the Japanese. To the U.S. technical planners, a single channel radio seemed very limited operationally, especially when jamming capabilities would block the channel. To not use jamming for the purpose of being able to listen to enemy communication for gathering information was of little value, since the U.S. code breakers could read Japanese encrypted messages and gathered valued information from that source. Therefore, communication jamming was not widespread in South Pacific as compared to radar jamming.

In outward appearance, Japanese radios were neat and modern looking. They added little to the weight of the aircraft, could be installed in the least amount of space, and required minimal special wiring. When the required concept of single frequency radios advanced to more sophisticated equipment, this was accomplished by making copies of foreign equipment, such as the higher power Liaison sets, where 100+ Watts of RF power were generated.

Early radar set dimensions were dictated by wavelength-related turned circuits and high voltage. To keep these outer dimensions of the units to a minimum, internal components were often bolted to the inner faces of their protective cases. This made replacement of some components very tedious. The very compact sets often overheated, since in an effort to achieve greater output from RF amplifier and modulator tubes they were operated close to their maximum ratings. This caused a concentration of heat that also shortened the life of adjacent components and the tubes. In an effort to cut down on overheating there was an engineering recommendation to operate radio equipment without outer covers, where possible, to allow convection to carry away the heat.

*Airborne Radar Technical Archives

Radio operator's position in the back seat of the newly restored Aichi M6A1 *Seiran*. At the right can be seen a portion of the Navy Type 96 Ku Mk.2 Command Radio. Other communication components to the left are the RDF Receiver, below which is the control box.

Japanese equipment did not employ auto-tune mechanisms, such as appeared on British TR-1143 or U.S. SCR-522 Command sets, or T-1154 or ART-13 Liaison sets. Usually the transmitter and receiver of Japanese Liaison sets were provided with two to four crystals installed before flight. These were connected into the oscillator circuits with a selector switch, and then the following stages were turned manually by the operator. On single-seat aircraft, only one transmitter and sometimes one receiver crystal-controlled channel was provided. Service equipment used by ground crews pretuned channels before flight when changes were needed.

During the latter half of the war, stockpiles of vacuum tubes and other materials needed in production of radio equipment were becoming exhausted. An effort was made to substitute critical materials with iron and wood whenever feasible. Even though most Japanese vacuum tubes were of American design, and were manufactured in Japan under license before the war, they tended to be less reliable because of material substitutions and process changes that could not be resolved by lengthy lab tests before they were pressed into production. Tube reliability was also affected because some were manufactured in smaller lots in smaller factories, and thus provided fewer units to select those meeting original specifications. Because of large wartime need for tubes for military equipment the acceptance limits had to be relaxed, leading to substandard performance in circuits, such as decreased sensitivity or spurious oscillations.

The two lists of aviation radios that follow cover the types most used by the Imperial Japanese Army and Navy, giving their type, with a degree of information felt appropriate to assist in identifying existing radios that may otherwise not be known. Where possible, the aircraft in which these were installed is given. These are not complete lists, but cover the major types of radio equipment that were in Japanese service. Variations in circuit, tube combinations, and layout may have occurred with or without any indication in the set nomenclature. Model or improvement numbers have not been included here where only minor variations are known to have

This is an Army Type 96 *Hi* Mk-2 Command Transmitter/Receiver, normally mounted in the aft gunner's compartment in two place aircraft. It was supplied with 6 receiver and 4 transmitter plug-in tuner assemblies to cover military frequency range. The transmitter had a relatively low output power. Also copied from TAIC Manual No. 1. *Courtesy of Todd Pederson*

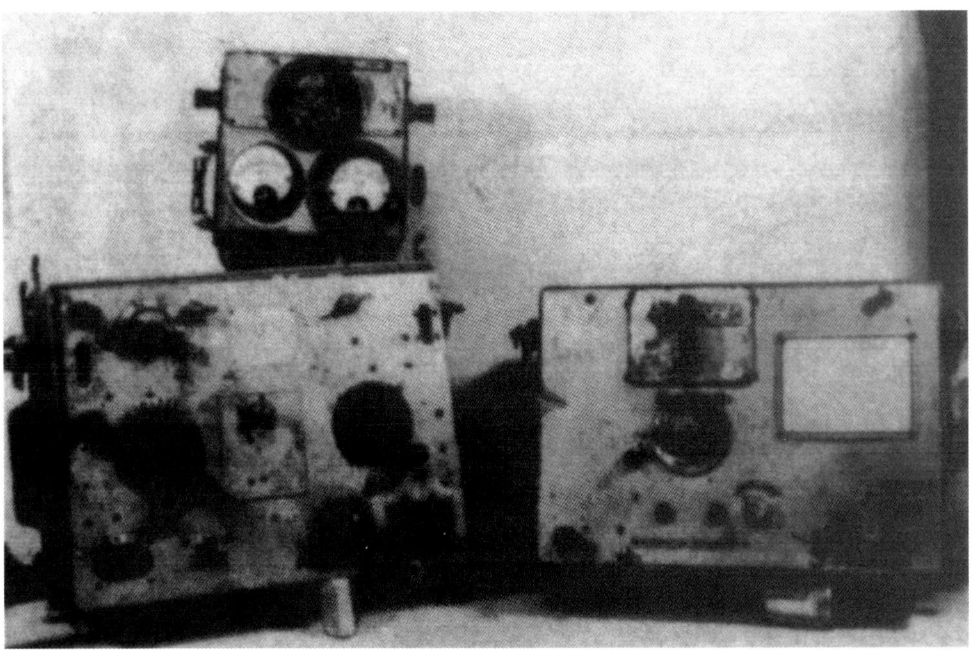

These components comprise the Army Type 94 *Hi* Mk.2 Command Radio. These were used in early twin-engined Army bombers that saw WWII service. Photo quality is poor because it is the only known image of this radio. Copied from TAIC Manual No.1. *Courtesy of Todd Pederson*

This is an Army Type 99 *Hi* Mk 1 Liaison/Command Radio. It was designed to eliminate the need for multiple plug-in assemblies and used only two transmitter and two receiver tuners. The transmitter plug-ins had coils with multiple taps connected to a patch panel. Flying leads to the oscillator circuit were plugged into the patch panels to tune the set to the required frequency. Flying leads were fragile, tangled easily with other wiring, and were easily broken, therefore this concept of patch panels did not last long. However, the set had respectable output power and served as Liaison Radio on large aircraft for some time. *Courtesy of Richard Lane.*

View of the Army Type 99 *Hi* Mk 1 radio transmitter with outer case removed. *Courtesy of Richard Lane.*

This lower portion of the Army Type 99 *Hi* Mk 1 is the receiver with outer case removed. This receiver is missing the tuner knob, right of center. *Courtesy of Richard Lane.*

occurred. As a reminder, the Type number identified the year in which the equipment was accepted as a military item, i.e. Type 96, year of Japan 2596 (1936). Command Radios were for Air-to-Air and Air-to-Ground communications, while Liaison Radios were designed for Long distance communication.

Designation List of Japanese Army Wireless Communication, Direction Finders, and Radar

Army Designation System

An identifier of Army aviation radio equipment is the character for *Hi* (飛), following the Type number. This character by itself means "to fly," but as the first character for "*Hikoki*" (飛行機) means airplane, which is considered the functional identifier in abbreviated form. Within this list when known, additional information is included to assist in identifying this equipment: (1) weight in pounds; (2) size in inches; (3) frequency range in megacycles; and (4) wattage, followed by remarks in parentheses.

Army Radio Equipment

Type 94 *Hi* Mk.2 (Command Radio)
(1) 36, (3) 1.4-7.9, (4) 15-20, (Mfg. Mitsubishi Denki, Tokyo Musen Denki, Nippon Musen Denshin Denwa. Used in Lily, Sally, Helen, Others.)

Type 96 *Hi* Mk.2 (Command Radio)
(1) 87, (2) 17 x 13 x 11, (3) 0.2-0.5, 1.5-7.5, (4) 8-10. (Mfg. Mitsubishi Denki, Tokyo Musen Denki, Nippon Musen, Denshin, Denwa. Used in: Helen, Lily, Dinah 1, Nate, Sonia, others. Obsolete by 1941 yet still in use on some aircraft as late as 1943.)

Type 96 *Hi* Mk.3 Model 2 (Liaison Radio)
(1) 15, (3) Trs. 2?, Rec. 5?, (4) 3, (Mfg. Toyo Tsushinki. Used in Lily, Sally, Nate, others. Obsolete by 1941 yet had a long service life.)

Type 99 *Hi* Mk.1 (or *Hi* Mk.1) (Command Radio)
(1) 74, (3) 2.5-18.0, (4) 30. (Mfg.: Tokyo Musen Denki, Nippon Musen, Kobe Seikosho. Used in: Sally, Helen, Dinah*. Standard radio for large aircraft, widely used.)
Note *: Dinah 3, 4, and Dinah 3 Modified Fighter used Type 99 *Hi* Mk.1 Modified.

Type 99 *Hi* Mk.2 (or *Hi* Mk.2 with modif.) (Command Radio)
(1) 42, (3) 1.5-15, (4) 18. (Mfg.: Tokyo Musen Denki, Tokyo Denki. Used in Lily 2, Dinah 2, Nick, others. Standard radio, well built but difficult to service. Unusual in having moisture retardant.)

This Army Type 99 *Hi* Mk-2 eliminated the plug-in problems associated with the Mk 1. These equipped newer Army twin engine and larger aircraft, as it was operated by a crew member other than the pilot. It provided 3 to 5 crystal controlled channels for the transmitter and a respectable power output. However, it was noted for being difficult to service in case of failures or battle damage. *Courtesy of Todd Pederson*

A component of the Army Type 99 *Hi* Mk.2 Command Radio is this antenna control box. Note that the brown color strongly resembles the interior color of the Kawasaki Ki-45 Kai Nick, a departure from the expected greens. *Courtesy of Todd Pederson*

The Army Type 99 *Hi* Mk 3 Radio Telephone was compact in design for use in single seat fighter aircraft. The transmitter was separate from the receiver, and both were about the same size. This is the cockpit of the Kawasaki Ki-61 Tony, showing the tunable receiver placed in front of the pilot. This small panel measures 8" x 7" x 5.5" deep. *80-G-192104*

This head-on view shows the face of the Army Type 99 *Hi* Mk.3 receiver. This is mounted in a panel removed from a Nakajima Ki.43 Oscar I when being evaluated at Eagle Farms, Brisbane. The square unit at the top center with a "U" bar handle is the frequency band unit that could be changed with others to give different frequencies.

This is the transmitter of the Army Type 99 *Hi* Mk.3 Command/Liaison Radio. It measures 7.5" x 8.5" x 15" deep and was mounted remotely in the rear of the aircraft. It provided one crystal controlled channel preset before flight. This transmitter is in the U.S. Air Force Museum collection. *Courtesy of Todd Pederson*

This composite view copied from TAIC Manual No.1 shows all the units that complete the Army Type 99 *Hi* Mk 3 Command/Liaison Radio.

Type 99 *Hi* Mk.3 (or *Hi* Mk.3 with modif.)
(Command/Liaison Radio)
(1) 46, (2) Trs. 6 x 7 x 8, Rec. 8 x 7 5-1/2, (3) Trs. 2.5-5.0, Rec. 1.5-6.7, (4) 8 Phone, 15 CW.(Mfg.: Toyo Tsushinki, Mitsubishi Denki, Nippon Denki, Tokyo Musen Denki. Used in: Tony, Tojo, Oscar, Sonia, Dinah, Frank, Ki-100, others. Primarily for Army fighters. Very compact.)

Type 99 *Hi* Mk.4 (or *Hi* Mk.4 with modif.) (Liaison Radio)
(1) 46, (2) 15 x 7-1/2 x 8-1/2, (3) 44.0-50.0, (4) 12. (Mfg.: Nippon Musen, Denshin Denwa, Kawanishi Kikai Seisakusho, Nippon Musen.
Used in: Sally, Lily 2, Helen Peggy. Had better I.F. tuning range and more flexibility than earlier models.)

Type 99 *Hi* Mk.5 (or *Hi* Mk.5) *Ko* & *Otsu*
(1) 119. (Mfg.: Adachi Denki K.K. Often interchanged with Type 99 Hi Mk.4. Not used as widespread. For Army fighters and Peggy. With this set the pilot had control over two bands for setting frequencies through a remote control unit. The pilot could then listen on either of both frequencies.)

Type 4 *Hi* Mk.3 (with modifications) (Command/Liaison Radio)
(Mfg.: Toyo Tsushinki. Used in Frank, Ki-100, others. An inferior version of the Type 99 *Hi* Mk.3.)

Hi Mk.1 Model 2 Radio Direction Finder (RDF)
(1) 24, (2) 17-1/2 x 7 x 9-1/2. (Mfg.: Nippon Musen Denko. Used in Sally, Lily 2, Helen, Peggy, others. Similar to German Telefunken Type EZ-2, P53N (AFN1 RH.I. System. Telefunken mfg RDFs of this type were Type *Te* and used in Lily 1 and Sally 1.)

Hi Mk.2 Radio Direction Finder (RDF)
(1) 60. (Used in Dinah 2, others. Smaller version of *Hi* Mk.1 RDF.)

Ta Ki-1 (Model 1 Radar) (Air Surveillance Vessel) (ASV)
(Used in Sally, Helen, Peggy, others. Standard Army radar for ASV and AI. Used TA Ki-1 altimeter box for low level bombing and torpedo attacks. Range of Detection: Aircraft - 20 km, 3,000 ton ship, 60 km, 1,000 ton ship, 30-35 km, 500 ton ship, 15-20 km, submarine 12-15 km.)

Ta Ki-5 ECM
(Small quantity built. Carried by Peggy.)

The Army Type 99 *Hi* Mk. 4 was the Army standard VHF Liaison set used widely throughout the war. VHF propagation characteristics limited its maximum range, and therefore it was not expected for long range liaison work. *Courtesy of Todd A. Pederson*

Junction Box

Dynamotor

Modulator

Transmitter

The Army Type 4 Mk 3 was a redesign of the Army Type 99 *Hi* Mk.3 to simplify production and use more available materials. It was packaged for ease of installation for primary use in multi-seat aircraft. Shown here are some of the major components that make up the Type 4. This grouping of 7 units is in the NASM collection. *Courtesy of NASM*

This poor quality photo for recognition purposes is only of the Army *Hi* Mk 1 Model 2 Radio Direction Finder, copied from TAIC Manual No.1. These German designed sets were mostly used in early wartime multi-engine aircraft. *Courtesy Todd Pederson*

Designation List of Japanese Navy Wireless Communication and Radio Wave Weapon For Aircraft, 1936-1945.

Navy Designation System

Within the Navy system of identifying electronic equipment, a standard numbering pattern was followed beginning with Type, followed by Mark Number, Name of Object, Model, and/or Modification. These terms are explained as follows:

Type: This identifies the year of acceptance as a military object. Example: Type 1 identifies the year of Japan 2601 (1941).

Mark *Ku*: Mark numbers in the case of radios is prefixed by *Ku* (空), the second character for the Navy designator "*Kokuki*" (航空機) meaning aircraft and implying airborne equipment. The number following identified the function intended for the equipment:

Ku Mk.1: Primarily for single seat aircraft.
Ku Mk.2: Primarily for two-seat aircraft.
Ku Mk.3: Primarily for three-seat aircraft.
Ku Mk.4: Primarily for multi-seat aircraft.
Ku Mk.5: Portable wireless telegraph able to be carried in aircraft.
Ku Mk.6: Radar.
Ku Mk.7: ECM (Electronic Counter Measure)
Ku Mk.8: Airborne radio for paratroopers.
Ku Mk.9: For rescue boats.

Name: This is generally self explanatory in describing the equipment. Wireless Telephone and Telegraph are early terms for radio, yet these terms are retained so as not to lose certain elements that radio alone does not adequately describe.

Model Number/or Modification: Model number differentiates equipment in the same year of acceptance and class having the same Type number. Modification is self explanatory.

Example: Type 96 *Ku* Mk.2 Wireless Telegraph *Kai*-3. The *Kai*-3 denotes the 3rd modification to wireless telegraph equipment designed for a two-seat aircraft, accepted in 1936.

Navy Radio Equipment

Navy Headquarters had both a long and short designation for the radio equipment. As the war progressed, the Ministry of Munitions applied their own shortened system. Under each long designation listed below is the (1) Navy Short Designation, and (2) Ministry of Munitions Short Designation. Where known, additional information is included to assist in identifying equipment: (3) weight in pounds; (4) size in inches; (5) frequency range in megacycles; and (6) wattage, followed by remarks in parenthesis.

Radios

Type 96 *Ku* Mk.1 Wireless Telephone (Command Radio)
(1) H1, (3) 38 Lbs, (4) 9-1/2 x 8 x 5-1/2 each unit, (5) 3.8-5.8, (6) 8-10. (1st year of the war in fighters, Mfg. by Oki Denki, Toyo Tsushinki. Used in Claude, Zeke, Rufe, others.)

Type 96 *Ku* Mk.2 Wireless Telegraph *Kai*-2 (Command Radio)
(1) N2 *Kai*-2, (2) *Fuji* 1, (3) 48, (4) 17 x 13 x 10-1/2, (5) .3-.5, 1.7-3.5, 5.0-10.0, (6) 27 CW, 9 voice. (Mfg.: Oki Denki, Nippon Musen. Used in Glen, Slim, Dave, Norm, Paul, Pete, Liz, Irving, Myrt, *Seiran,* Dinah (Navy), Judy, Val, others.)

This is a rare view of a Zero, in that it shows the Navy Type 96 *Ku* Mk. 1 Transmitter and Receiver installed. This Zero was downed during the Pearl Harbor attack. This radio is unusual for a single seat fighter in that both the Transmitter and Receiver were designed to fit in the cockpit and be accessible to the pilot. On the port side of the cockpit was the power supply from two Dynamotor. This installation was similar to that found in the A5M Claude. *80-G-22158*

Above: Often found in two seat navy aircraft was this Navy Type 96 *Ku* Mk. 2 that was standard Command/Liaison Radio equipment through most of the war. These were operated by the rear seat occupant of two seat aircraft. These were regarded as "state of the art" at the time of their introduction in 1936 and for a time afterwards. Below: Top view with access panels open of the Navy Type 96 *Ku* Mk.2 Radio. *Courtesy of Todd Pederson*

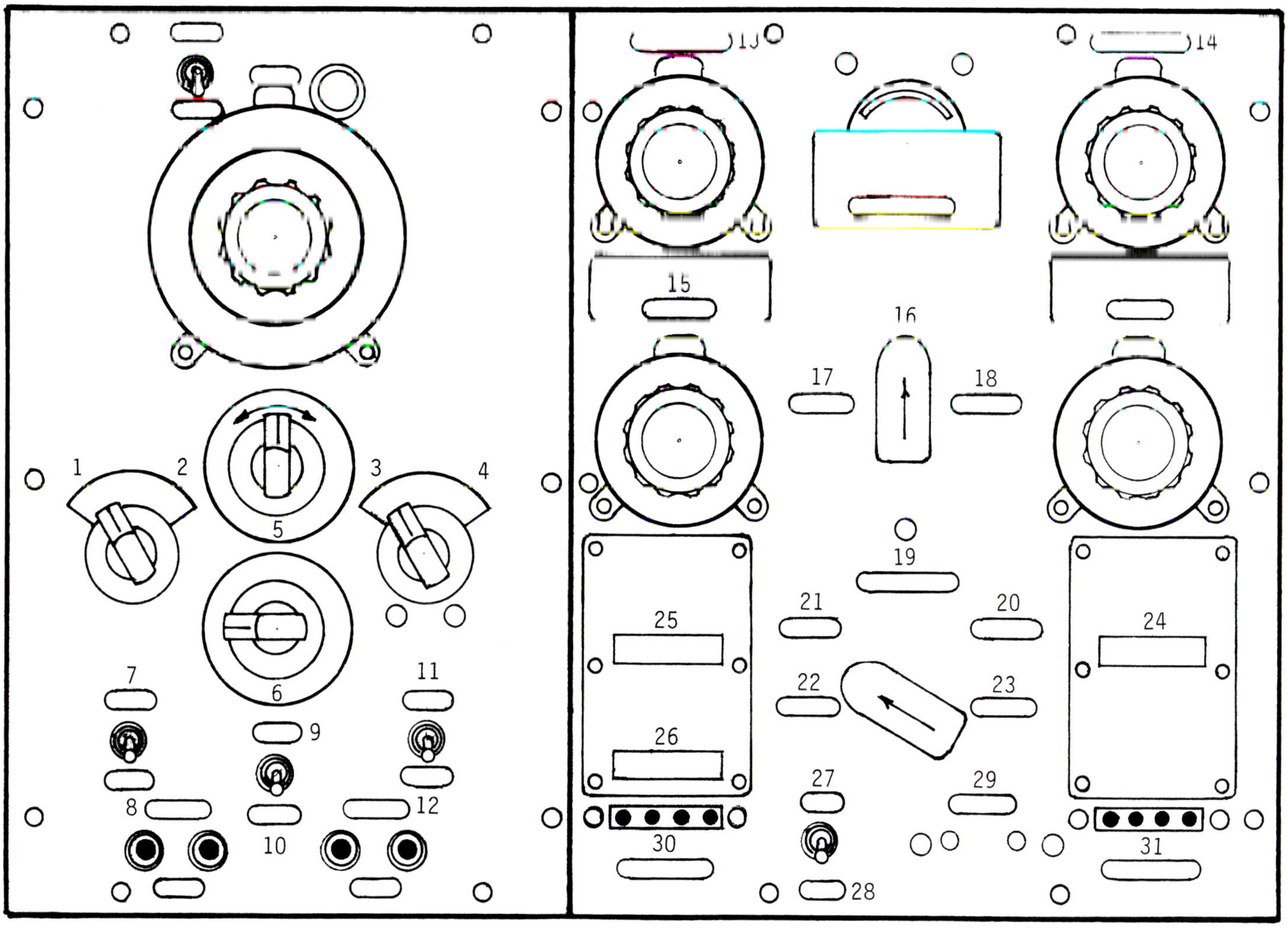

This graphic of the Navy Type 96 *Ku* Mk 2 front panel illustrates typical controls that were found on most Japanese WWII radios. Functions of many are long forgotten and therefore recorded here for posterity and technical reference.

1. Telephone
2. Telegraph
3. Short Wave
4. Long Wave
5. Modulation
6. Volume Control
7. Shift to Front Cockpit
8. Shift to Rear Cockpit
9. AVC
10. Off
11. Transmitting and Receiving
12. Transmitting only
13. Antenna Tuning
14. Antenna Tuning
15a. Oscillator Tuning, Short Wave
15b. Oscillator Tuning, Long Wave
16. Off
17. Transmitting and Receiving
18. Receiving
19. Wave switch
20. Short Wave 1
21. Short Wave 2
22. Short Wave 3
23. Long Wave
24. Short Wave Crystal 1
25. Short Wave Crystal 2
26. Short Wave Crystal 3
27. Tuning
28. Transmitting and Receiving
29. Telegraph Key Socket
30. Rear Cockpit Microphone Socket
31. Front Cockpit Microphone Socket
32. Rear Cockpit Headphone Socket
33. Front Cockpit Headphone Socket
34. Antenna Current Meter
35. Receiver Tuning

Seen here is a typical installation of the Navy Type 96 *Ku* Mk.2. This unit is installed in the rear cockpit of the Aichi A6M1 Seiran.

Shown here are most of the units to make a complete Navy Type 96 *Ku* Mk.2 Kai 2 Command Radio.

This is a closer view of the antenna tuner as part of the Navy Type 96 *Ku* Mk.2 Kai 2. *Courtesy of Todd Pederson*

Resting on the narrow ledge on the starboard side of *Seiran*'s rear cockpit is the CW key. CW (carrier wave) transmission with a telegrapher's key (for Morse code) was the most reliable form of communication during the War. This was used with the Navy Type 96 Ku Mk.2 installed in this airplane.

Another version of the Navy Type 96 *Ku* Mk 2 was an "Improvement 2," or Kai-2 Command Radio. It replaced two of the crystal holders in the transmitter with a continuously tunable oscillator-wavemeter. ***Courtesy of Todd Pederson***

Type 96 *Ku* Mk.3 Wireless Telegraph (Command Liaison Radio)
(1) H3, (3) 40, (4) Rec. & Tns. 18-1/2 x 14 x 10, Rec. only 14 x 10 x 7-1/2, (5) 0.2-0.5, 4.5-11.2, (6) 75 Hi, 8 Lo. (Mfg.: Oki Denki, Nippon, Denki, Kaigun Kokusho, Toyo Tsushinki, Tokyo Denki, Adachi Denki.
Used in Norm, Paul, Pete, Lorna, Betty, Jake, Val, others.)

Type 96 *Ku* Mk.4 Wireless Telegraph (Command Radio)
(1) H4, (3) 167, (5) 0.3-15.0, (6) 175 CW. (Mfg.: Tokyo Denki. Receiver is similar to Type 96 *Ku* Mk.3 but with tone tube omitted. Used in Betty, Nell, Mavis, Emily, others.)

Type 98 Ku Mk.2 Command Radio.
No information or photo available, but known to have been used aboard Val.

Type 98 *Ku* Mk.4 Command Radio Inter-formation Telephone
(1) U4, (2) *Hagi* 2, (3) 30, (4) 16 x 8 x 11, (5) 29.5 52.5, (6) 12 (voice) (Mfg.: Oki Denki, Nippon Musen, Denshin Denwa. Used in Betty, Nell, Rita, Emily, Mavis, others. Obsolete but retained.)

Type 1 *Ku* Mk.3 Liaison and Command Radio Inter-formation Telephone *Kai*-1
(1) U3 *Kai*-1, (2) *Hagi* 1, (3) 25, (4) 14 x 6 x 6, (5) 30.0-45.0, (6) 5 (voice).(Mfg.: Nippon Musen. Used in Alf, Laura, Irving, Jack, Kate, Jill, Frances, Paul, Grace, Jake, Val, Judy, others. Standard VHF throughout the war.

This is a Navy Type 96 Ku Mk 3 Wireless Telegraph Liaison Radio used in mufti-seat aircraft. It used a higher power transmitter (75 w) for long-range work. The receiver is the top unit, transmitter on the bottom. This only known photo of this radio is copied from TAIC Report #1, Japanese Aircraft Performance and Characteristics. ***Courtesy of Todd Pederson***

Large Navy aircraft were equipped with this Navy Type 96 *Ku* Mk 4 Wireless Telegraph. It had a higher power transmitter output (175 w) than the Mk.3, yet the receiver was very similar to the Mk 3. Constructed in two sections, the Receiver is on the top, bolted to the Transmitter below. The knobs at the sides are for shock proof bungee mounting. ***Courtesy of Todd Pederson***

Navy Type 98 *Ku* Mk.4 Command Radio was a VHF set and a forerunner of the Navy Type 1 Ku Mk.3 set. Other components needed for this set are the Dynamotor, telegraph key, headset, and microphone. ***Courtesy of Richard Lane***

This top view of the Navy Type 1 Ku Mk.3 *Kai* 2 Liaison and Command Radio has top covers removed to show interior configuration. *Courtesy of Todd Pederson*

This Type 1 *Ku* Mk.3 Kai 2 (second modification) Liaison and Command Radio is surviving equipment that came installed in the NASM Kugisho P1Y1 Frances. This is a VHF radio considered standard equipment throughout the war. *Courtesy of Todd Pederson*

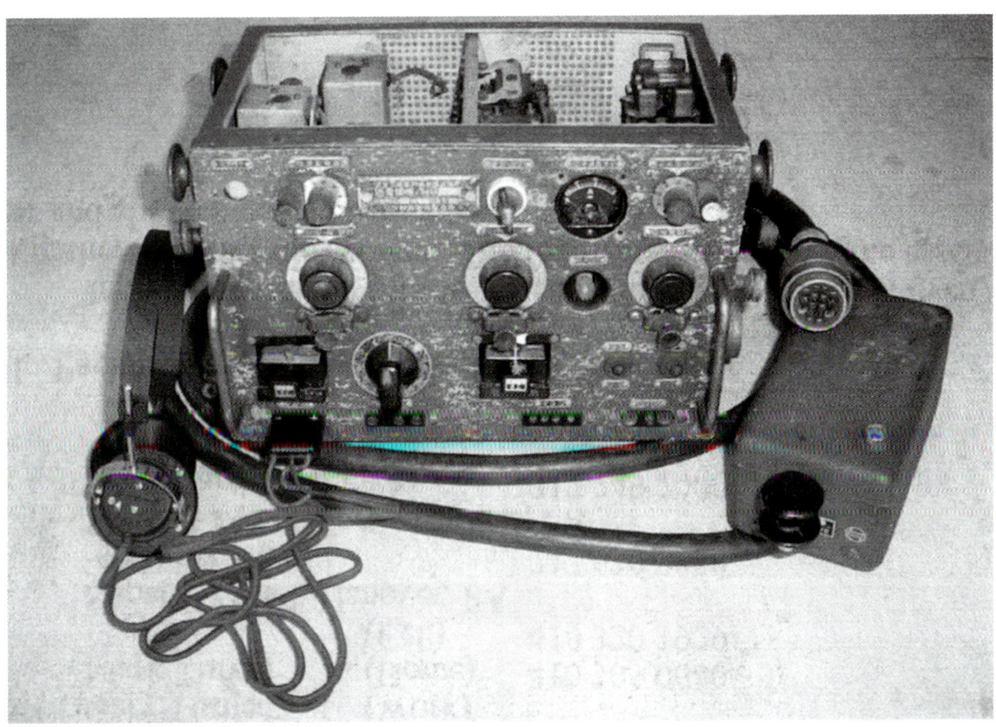

Another radio of the same Navy Type 1 *Ku* Mk.3 Kai 2 shows the unit with headset (left) and telegraph key (right). Other components needed are a Dynamotor, control box, auxiliary volume control, and frequency adjustment tool. *Courtesy of Todd Pederson*

Type 2 *Ku* Command Radio Inter-formation Telephone
(1) UP (For all types of aircraft.)

Type 2 *Ku* Command Radio Inter-formation Telephone
(1) UP *Ko* (Same as UP for 24 v.)

Type 2 *Ku* Mk.3 Wireless Telegraph *Kai*-1 (Liaison Radio)
(1) N3 *Kai*-1, (2) *Fuji* 5, (3) 172, (4) 21 x 15 x 12,(5) 0.3-20.0, (6) 200 CW, 60 voice. (Mfg.: Nippon Musen. Large unit. Used in Paul, Jake, Kate, Jill, Frances, Betty, Nell, Emily, Mavis Irving, Myrt, Grace, Rita, *Keiun*, others.)

Type 3 *Ku* Mk.1 Wireless Telephone *Kai*-3 (Command and Liaison Radio)
(1) N1 *Kai*-3, (2) *Sumire* 1 *Kai*-3, (3) 65, (4) 17 x 8 x 8, (5) 5.0-10.0, (6) 75 CW, 25 voice, (Mfg.: Toyo Tsushinki, Hayakawa Denki Kogyo. Used in Judy, Rufe, Rex, Zeke, George, Jack, Sam, Betty, Liz, *Shinden*, others.)

Type 3 *Ku* Mk.1 Receiver
(1) N1-P, (2) *Sumire* 5 (For kamikaze attack aircraft. To be replaced by P1-R.)

Type 3 *Ku* Mk.5 Wireless Telegraph
(1) GS, (3) 150, (4) Trs. 27 x 21 x 15, Rec. 15 x 12 x 12, (5) Trn. 2.5-20.0, Rec. 0.3-20.0.(6) 75. (Mfg.: Oki Denki. Used in Emily, Mavis, Betty, Nell, others.)

18-*Shi* Experimental *Ku* Mk.3 Wireless Telegraph *Kai*-1
(1) P3 *Kai*-1 (Mfg.: Numazu Naval Arsenal. For two and three-seat aircraft.)

18-*Shi* Experimental *Ku* Mk.5 Wireless Telegraph
(1) AGS (Mfg.: Numazu Naval Arsenal. Same as GS.)

18-*Shi* Experimental *Ku* Mk.8 Wireless Telegraph
(1) P8 (For paratroops and spotter stations.)

19-*Shi* Experimental *Ku* Mk.1 Wireless Telephone
(1) P1 (For fighters.)

19-*Shi* Experimental *Ku* Mk.1 Receiver
(1) P1-R (For kamikaze attack aircraft.)

19-*Shi* Experimental *Ku* Mk.1 Receiver *Ko*
(1) P1-R *Ko*, (2) *Sumire* 5 *Ko* (For kamikaze attack aircraft. For 24 v.)

19-*Shi* Experimental *Ku* Mk.3 Wireless Telegraph
(1) R3, (2) *Fuji* 8 (Mfg.: Numazu Naval Arsenal. For two and three-seat aircraft.)

19-*Shi* Experimental *Ku* Mk.3 Wireless Telegraph *Ko*
(1) R3 *Ko*, (2) *Fuji* 8 *Ko* (For two and three-seat aircraft. For 24 v.)

19-*Shi* Experimental *Ku* Mk.4 Wireless Telegraph
(1) R4, (2) *Fuji* 10 (Multi-seat aircraft version of R3.)

19-*Shi* Experimental *Ku* Mk.4 Wireless Telegraph *Ko*
(1) R4 *Ko*, (2) *Fuji* 10 *Ko* (Same as R4 with 24 v.)

19-*Shi* Experimental *Ku* Mk.9 Transmitter
(1) P9, (2) *Fuji* 9 (For rescue boat.)

The Navy Type 2 *Ku* Mk.3 Liaison and Command Radio was a large unit for long range and suitable for larger aircraft. It consists of three basic units shown separately on page 89. *Courtesy of Richard Lane*

This top unit of the Navy Type 2 *Ku* Mk.3 radio is the receiver, shown exposed without its top cover, to which the sides are also attached. It has a self contained small dynamotor that is at the left. *Courtesy of Richard Lane*

The middle unit of the Navy Type 2 *Ku* Mk.3 radio is the tuning unit, which contained the two RF power vacuum tubes extending from the unit below. When fully assembled, access to these tubes for replacements is through the outside grill acting as a door. *Courtesy of Richard Lane*

This is the bottom unit of three, which is the transmitter that utilizes the two RF power vacuum tubes extending above the unit. *Courtesy of Richard Lane.*

One of the better known Japanese airborne radios was this Navy Type 3 *Ku* Mk.1 Liaison and Command Radio found in many Japanese Navy aircraft. It differed considerably from earlier sets in that the transmitter and receiver were a single unit made to mount away from the cockpit and operated by the pilot through a control box. It had provision for crystal controls of both transmitter and receiver. In this original Japanese photo from an operator's manual are the dynamotor at left and pilot's helmet with earphones at right. Pilot's control unit shown below is to the right of the main unit above. *Courtesy of Todd Pederson*

Here are two remote control heads for the Navy Type 3 *Ku* Mk.1 Liaison and Command Radio that are located in the cockpit, usually in single place aircraft. This allows the large major unit to be located in the rear of the aircraft. An early Mk.1 is at left, and a later Mk.1 Kai is at right. *Courtesy of Todd Pederson*

Radio Direction Finders (DF)

Type 0 *Ku* Mk.4 Radio Direction Finder

(1) F4, (2) *Ran* 4, (3) 75, (4) Varies, (5) 0.165 1.2. (Mfg.: Nippon Musen. Used in Betty, Emily, Mavis, Rita, Liz, Tabby, Frances, Nell, others.)

Type 1 *Ku* Mk.3 Radio Direction Finder

(1) FF, (2) *Ran* 1, (3) 15, (4) 18 x 9 x 6, (5) 0.17-1.2. (Mfg.: Tokyo

Denki, Nippon Denki. Used in Zeke 32, 52, Kate, Val, Sam, Myrt, Susie, Jake, Judy, Jill, Babs, Glen, Slim, Frances, Jill, Grace, Irving, Rex, Paul, Norm, *Seiran*, *Keiun*, Jack, George, Emily, Mavis, Betty, Nell, Alf, Laura, others. Std. for Navy throughout the war. Copy of U.S. Fairchild Krusei RDF that was imported by the Navy in large quantities and used as Type *Ku* RDF.)

19-*Shi Ku* Experimental Radio Direction Finder

(1) FP, (2) *Ran* 6 (For all sizes of aircraft.)

19-*Shi Ku* Experimental Radio Direction Finder *Ko*

(1) FP *Ko*, (2) *Ran* 6 *Ko* (Same as FP for 24 v.)

Interphone

18-*Shi* Experimental Interphone

(1) A3, (2) *Sakura* 3 (For multi-seat aircraft, but used specifically in Irving, Jill, Frances.)

18-*Shi* Experimental Interphone *Ko*

(1) A3 *Ko*, (2) *Sakura* 3 *Ko* (For multi-seat aircraft for 24 v.)

This unexciting view, yet very rare, is of the receiver unit for the Navy Type 0 *Ku* Mk.4 Radio Direction Finder. It is shown here installed in the aft section of the Kugisho P1Y1 Frances belonging to NASM. The mechanical linkages are to hand cranks at the RDF Control Box. They tune the receiver to the station to be homed to and operate the sense circuitry. *Courtesy of Z.I. Szewezyk*

This very crude hand crank for the loop antenna for the Type 0 *Ku* RDF is located in the gunner/radio operator compartment of the Kugisho P1Y1 Frances. As the loop antenna is repositioned as to its relationship with the radio signal being received its signal strength varies. When the signal is the weakest, the plane of the loop is perpendicular to the station. *Courtesy of Todd Pederson*

Also in the rear compartment of Frances on the port side is this controlling device. It is part of the Type 0 *Ku* RDF used to determine which of the two headings 180 degrees apart will lead to a station. A sense-omnidirectional antenna is switched in to combine its signal with the loop antenna signal in a proper phase. The resulting pattern is a cardioid which has only one null, and therefore determines the correct direction to the station.

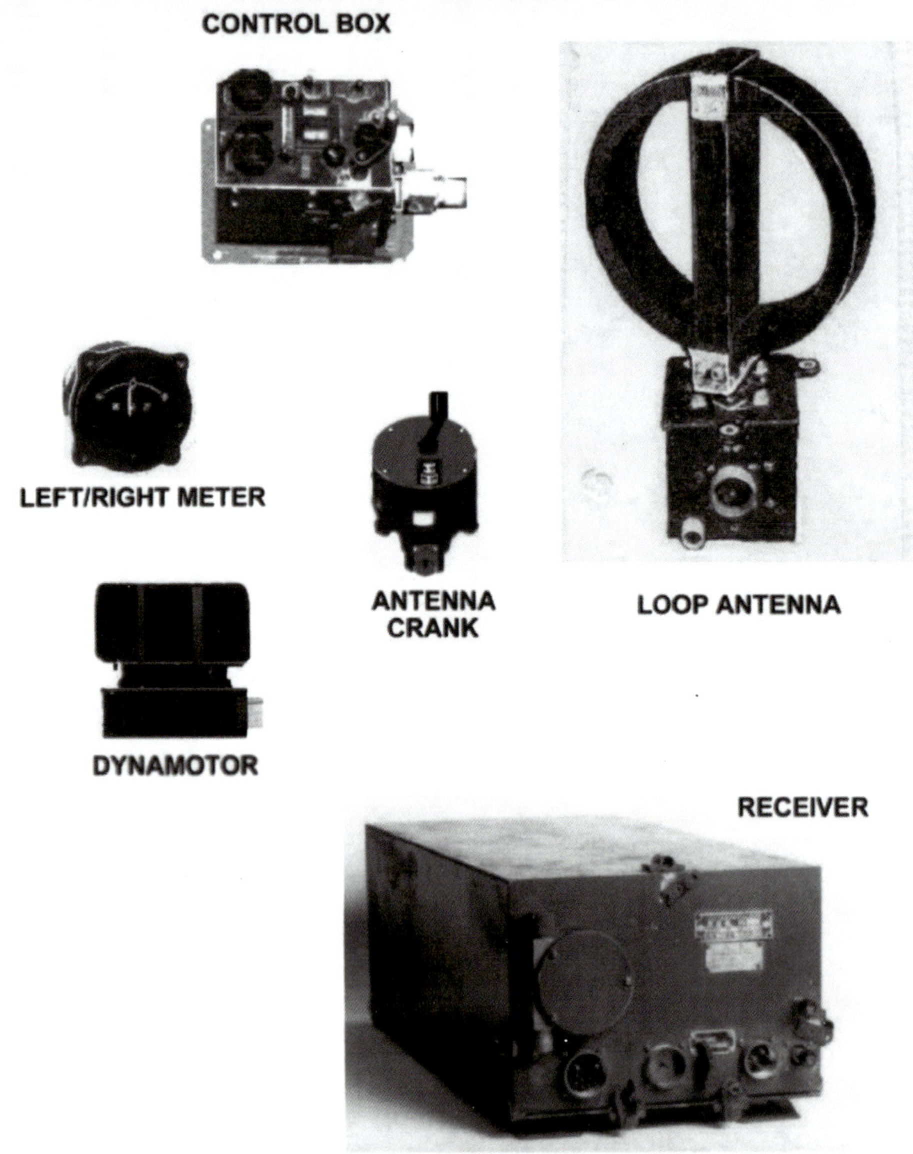

Ku 3 RDF

Except for being identified as a Radio Direction Finder, this equipment carried the same Navy Type 1 *Ku* Mk.3 designation as the Liaison and Command Radio, often causing confusion. This radio system was recovered from a downed Aichi E16A1 Paul by TAIU teams. It appears to be a copy of the U.S. built Fairchild Kruesi Radio Compass.

Radar

Type 3 *Ku* Mk.6 Wireless Telegraph Model 4 *Kai*-3
Function: ASV (Air-to-Surface Vessel)
(1) H6 *Kai*-3, (2) *Kaze* 1 *Kai*-3, (3) 240 lbs.(Mfg.: Nippon Musen, Kawanishi Kikai Seisakusho. Used in Frances, Irving, Nell, Betty, Lorna, Kate, Jill, Jake. Navy Std., similar to U.S. SCR 521 or British ASV Mk. II equipment. Range:10,000 ton ship, 51 km at 304 m (10,000 ft.), 30 km at 60 m (2,000 ft). True bearings were difficult to obtain and hold.)

Type 4 *Ku* Mk.6 Wireless Telegraph Model 4
Function: ASV (Air to Surface Vessel)
(1) FM1, (2) FM1, (3) 120 lbs. (Mfg.: Tokyo Denki. Used in Frances, Betty, Jill, Lorna, Irving, Jake, Kate, Nell 23. Had forward Yagi antenna and two waist mounted antenna.)

18-*Shi Ku* Mk.6 Wireless Telegraph
(1) FD2, (2) *Kaze* 13 *Kai*-1 (Mfg.: Tokyo Denki. For night fighters.)

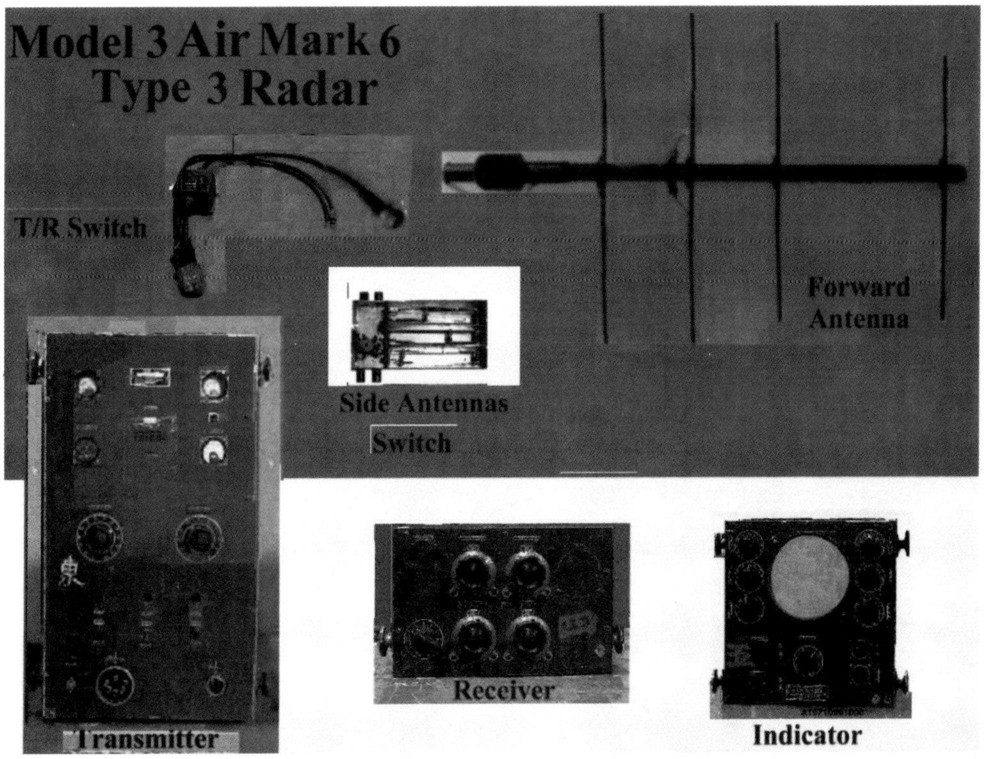

These three main radar units are components of the Navy Type 3 *Ku* Mk 6 Wireless Telegraph Model 4 Kai-3, commonly known as H-6 radar and in U.S. as Air Mk-6. The reason for this "Wireless Telegraph" nomenclature attached to these units was for deception of Japan having radar if captured by the enemy. This equipment was roughly equivalent to the US SCR 521 or British ASV Mk. II but much harder to operate. Antenna arrangements varied with the type of aircraft on which installed. This antenna configuration was used on the Nakajima J1N1-S Irving. *Courtesy of Z.I. Szewezyk*

This photo shows the Navy Type 3 *Ku* Mk 6 Wireless Telegraph Model 4 Kai-3 installed in the rear cockpit of a Nakajima B5N2 Kate captured at Saipan Island. The H-6 Indicator with 5 inch CRT rests below the receiver. To the right on the floor is the transmitter. The aluminum box above is American installed equipment. *80-G-122975*

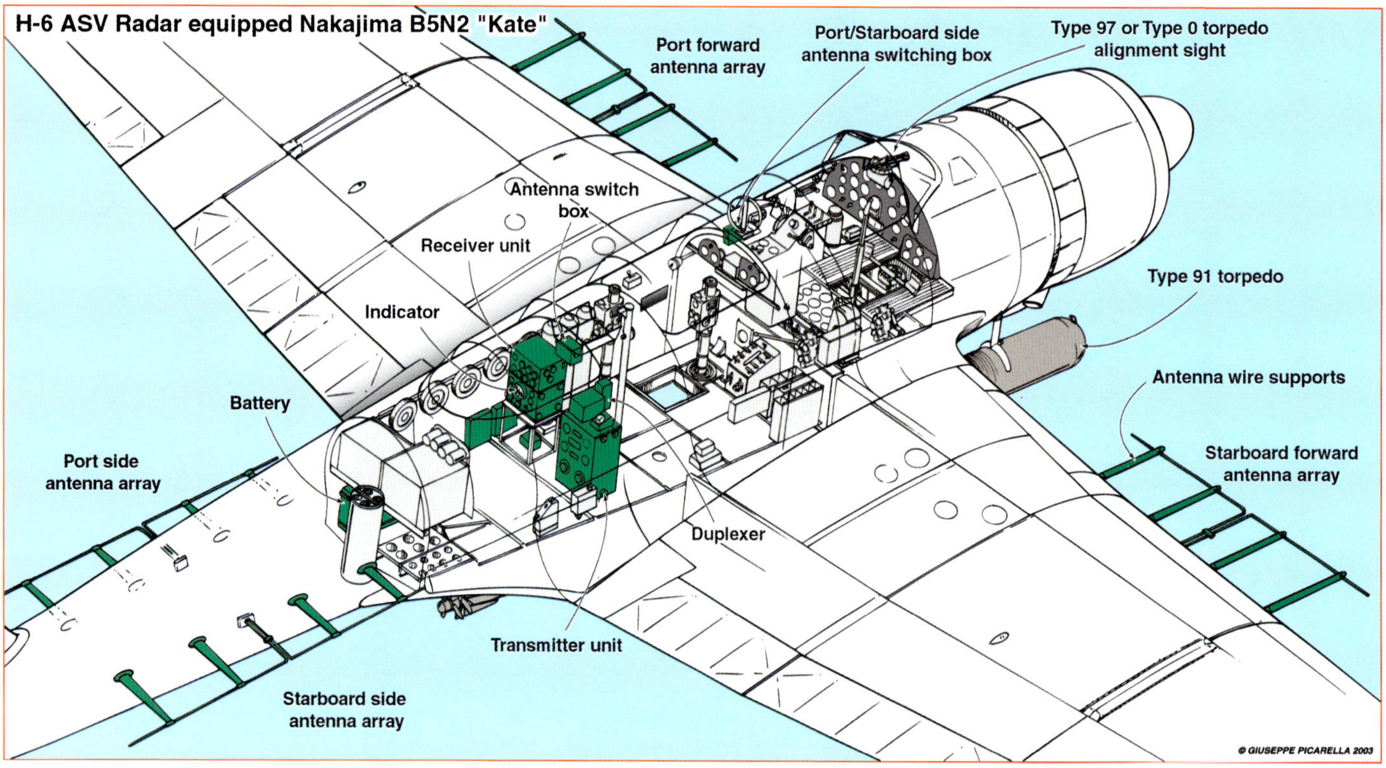

This drawing shows an installation of a 3 beam radar in a Nakajima Kate. This represents a Navy Type 3 *Ku* Mk 6 Wireless Telephone Model 4. The Forward, Port, and Starboard antenna arrays could be used one at a time. No other means were provided to measure direction to the target. ***Courtesy of Giuseppe Picarella***

19-*Shi Ku* Experimental Radar (Mk. 1) Model 13
(1) FK4 (For large aircraft.)

19-*Shi Ku* Experimental Radar (Mk. 1) Model 13 *Ko*
(1) FK4 *Ko* (For large aircraft for 24 v.)

19-*Shi Ku* Experimental Radar (Mk. 1) Model 11
(1) N6, (2) N6 (For small aircraft.)

19-*Shi Ku* Experimental Radar (Mk. 1) Model 12
(1) FK3, (2) FK3 (Mfg.: Kawanishi Kikai Seisakusho. For small aircraft.)

19-*Shi Ku* Experimental Radar (Mk. 2) Model 11
(1) *Gyoku* 3 (Mfg.: Nippon Musen. For night fighters.)

19-*Shi Ku* Experimental Radar (Mk. 3) Model 30.
(1) Mk. 51, microwave (10 cm) equivalent to H2S.

ECM (Electronic Counter Measures)
ECM *Kai*-4
(1) F27 (For large aircraft. Converted from marine ECM.)

Type 2 *Ku* Mk.7 Wireless Telegraph Model 2 *Kai*-1
(1) FTB *Kai*-1, (2) *Kumo* 4 *Kai*-1 Radar-warning receiver. (Seen at NASM as 1973-154).

Type 2 *Ku* Mk.7 Wireless Telegraph Model 2 *Kai-Ko*
(1) FTB *Kai*-1 *Ko*, (2) *Kumo* 4 *Kai*-1 *Ko* (For all sizes of aircraft with 24 v.)

Type 2 *Ku* Mk.7 Wireless Telegraph Model 3
(1) FTC, (2) *Kumo* 5 (For all sizes of aircraft.)

Type 2 *Ku* Mk.7 Wireless Telegraph Model 3 *Ko*
(1) FTC *Ko*, (2) *Kumo* 5 *Ko* (Same as FTC with 24 v.)

Note: The term *wireless telegraph* was used for marking Radar and ECM equipment carried in operational aircraft to help enforce secrecy as to purpose if captured by the enemy.

IFF (Identification Friend or Foe)
Experimental Type 5 *Ku* IFF *Kai*-1
(1) M13 *Kai*-1, (2) *Kiri* 2

Experimental Type *Ku* IFF *Kai*-1 *Ko*
(1) M13 *Kai*-1 *Ko*, (2) *Kiri* 2 *Ko*

Receiver Indicator for 19-*Shi Ku* Experimental Radar Mk.1 Model 12 (FK3).

This is the Receiver (top) and Transmitter (bottom) of the 19 *Shi* Ku Experimental Radar Mk.1 Model 12. When compared with the Type 3 Ku Mk.6 Model 4, this shows tremendous progress in reducing the size to about one half of the earlier type. Operating complexity was improved by reducing 25 controls to about 10, making the set useable by average radiomen.

Transmitter for 19-*Shi Ku* Experimental Radar Mk.1 Model 12. (FK3).

Instrument Landing System (ILS)
Type 4 Wireless Landing Receiver
(1) *Kiri* 3, (2) *Kiri* 3

Source: *General View of Aeronautical Engineering* (1955), chapter authored by Capt. Iwao Arisaka, supplemented by TAIC Manual No. 1, *Japanese Aircraft*, Radio Section, June 1945.

Associated Equipment

Each crew position in multi-place aircraft had an intercom box. They had a volume control and a receptacle into which the microphone and headsets were connected. ***Courtesy of Todd Pederson***

This view of the intercom box shows the headset (right) and the microphone plugged into the unit. Note the finger grip notches on the plug. ***Courtesy of Todd Pederson***

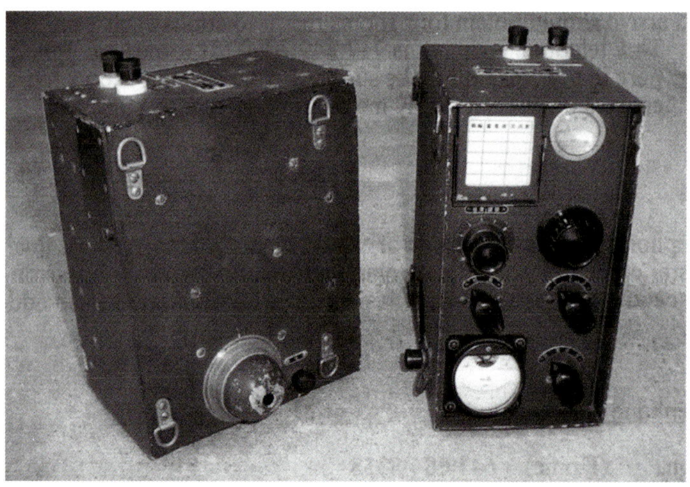

Navy Type 99 Radio Wave Measuring Apparatus. ***Courtesy of Todd Pederson***

Nomenclature identifies this unit as a Navy Type 2 *Ku* 7 Model 2 Radar Detector. Technical drawings of the Ohka 22 list a device of this description carried on board. Its purpose was probably to alert the pilot of detection and to take evasive action. It measures 9.5" x 15.75".

Above: CW, or carrier wave transmission with telegrapher key, like the one shown here, was the most reliable form of communication during the war because of its long range capability. Right: Loop antenna for Radio Direction Finding equipment. (Above and right) *Courtesy of Todd Pederson* Below: This Nakajima B5N2 Kate is equipped with a Navy Type 3 *Ku* Mk 6 Wireless Telegraph Model 4 Kai-3, which was actually airborne radar. This is the forward looking antenna attached to the wing leading edge. *80-G-122969*

Side view of Kate showing the side-looking radar antenna. The shadow on the floor reveals the presence of the antenna on the port side. ***80-G-122970***

This is the Yagi antenna mounted on the nose of the Nakajima J1N1-S Irving belonging to NASM. This airplane was one of five Irvings equipped with the 18-*Shi Ku* Mk.6, also called FD-2 radar, but not currently installed, as location within the aircraft was not positively identified.

Right: This is the side-looking radar on an Irving found on Tinian in August 1944. The fuselage skin acts as a reflector generating "image" radiator for this type installation. The tape measure held by the GI gives a false impression of the antenna design. *80-G-167821* Below: Close-up top view of an Irving nose, lying on its side with Yagi antenna. Note the 20-mm gun barrel protruding through the wooden nose block, a nonstandard arrangement variation of the Irvings of the 321st *Kokutai* found on Tinian. It is most likely an air-to-surface, sea-search radar since this is more aligned with the mission of the 321st. *80-G-167829*

4

Aerial Cameras

by
Richard A. Lane

The Japanese air forces used a variety of aerial cameras produced during the Pacific War. This section will not attempt to cover all types of aerial cameras produced for the Japanese military. Instead, it will focus on the principal cameras used during that war. Background material on aerial cameras is somewhat limited. The best source appears to be a study produced by the U.S. Army Air Forces at Wright Field, Ohio, in November 1946. This study is entitled *Japanese Photo Intelligence and Equipment in World War II* and contains text and black and white photos of the cameras, most of which have been replaced here with color images of the same type equipment in a private collection. Several Japanese Publications (i.e. Maru Mechanic) have also illustrated these cameras (although the text is in Japanese).

Army Cameras

The Type 99 Small Aerial Camera was probably the smallest Japanese camera of this type. On the camera case it is identified as the GSK-99. This camera took 10 to 11 exposures on ordinary 120 roll film, giving an image size of 6 cm x 6 cm. It weighed approximately 1.35 kg (3 lb), and its largest dimension was 20 cm (just under 8 in). The camera had an automatic film transport mechanism, and was wound by a crank on the bottom of the camera. Two extra backs were provided and could be quickly placed on the camera during operation. Shutter speeds ranged from 1/100 - 1/250 and 1/500 (or 1/100, 1/200, and 1/400) of a second. The Type 99 was made by *Konishi-Roku* (Konica), *Tokyo Kogaku* (Toko, later Topcon Camera), and possibly *Nippon Kogaku* (Nikko, later Nikon Camera). The camera

Japanese aerial cameras of the Pacific War period were made with the precision of world class counterparts. Shown here is the Army Type 99 Small Aerial Camera, which uses 120 roll film in the two film backs, shown along with three filters. ***All photos this chapter courtesy Richard A. Lane unless otherwise noted.***

This top view of the Army Type 99 Small Aerial Camera shows the ruggedness of not only the body, but that of the handles, as well. The optical viewfinder is the black object. Some Army Type 99 cameras, which on their label were identified as GSK-99 cameras, were equipped with similar viewfinders with no optical elements.

Army Type 96 (SK-96) Small Aerial Camera was one of the most sophisticated aerial cameras of WWII, yet hardly recognized as such because of bias. The maker of this camera was *Nippon Kogaku Kogyo K.K.* (Nikko).

illustrated has a fitted carrying case containing three colored filters and two spare film backs. Two supplementary lenses for portrait photography were usually included. This model has a serial number of 100, and *Tokyo Kogaku* (Toko) produced this camera in January 1942. It is equipped with a Toko 75 mm f3.5 lens. This camera was primarily used for hand-held oblique photography.

The Type 96 (also SK-96) Small Aerial Camera may have been the most sophisticated camera made for the JAAF. Its service life was relatively short, being replaced by the Type 100 (SK-100) Small Aerial Camera that was produced in far greater numbers. According to Air Technical Intelligence Review No. F-IR-87-RE, dated November 1946, titled *Japanese Photo Intelligence & Equipment in World War II*, the Type 96 suffered from reliability problems during operation that called for its replacement. The Type 96 used heated elements in the camera body and lens. Additionally, it was equipped with two light projectors for recording the tilt of the camera (in degrees) onto the film. It used 18 cm roll or sheet film. The JAAF used this camera for precision photo mapping, in addition to oblique strike photography. The camera could be used hand-held or mounted in an aircraft. Dimensionally, the camera is about the same size as the Type 100, although it weighs considerably more. The model illustrated here was made by *Nippon Kogaku Kogyo K.K.* (also written Nikko) in August 1941, and was equipped with a Nikkor 179 mm f4.5 lens. The camera's fitted carrying case holds four lens filters, four plate film backs, one roll film back, an optical view finder, two bubble levels, four canvas film packs, and a focusing screen. This model may have also been made by Konica and possibly others.

The Type 100 (SK-100) Small Aerial Camera was probably the most produced and used JAAF camera for hand held oblique

Army Type 96 Small Aerial Camera with accessories.

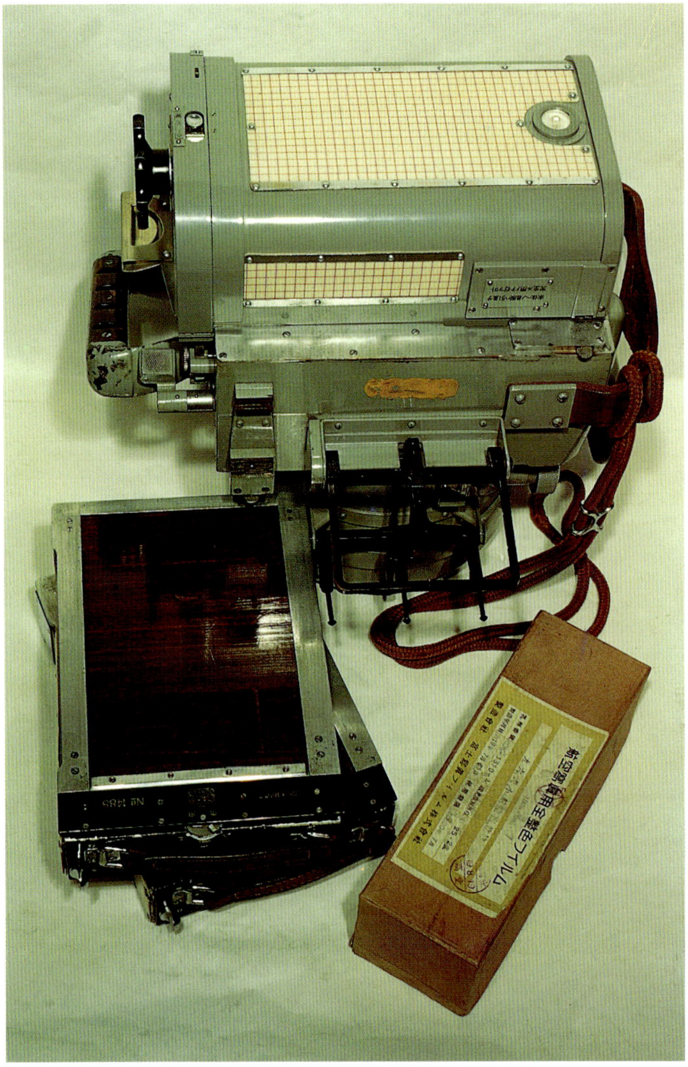

Army Type 96 Camera Small Aerial Camera with plate film holders and boxed roll film.

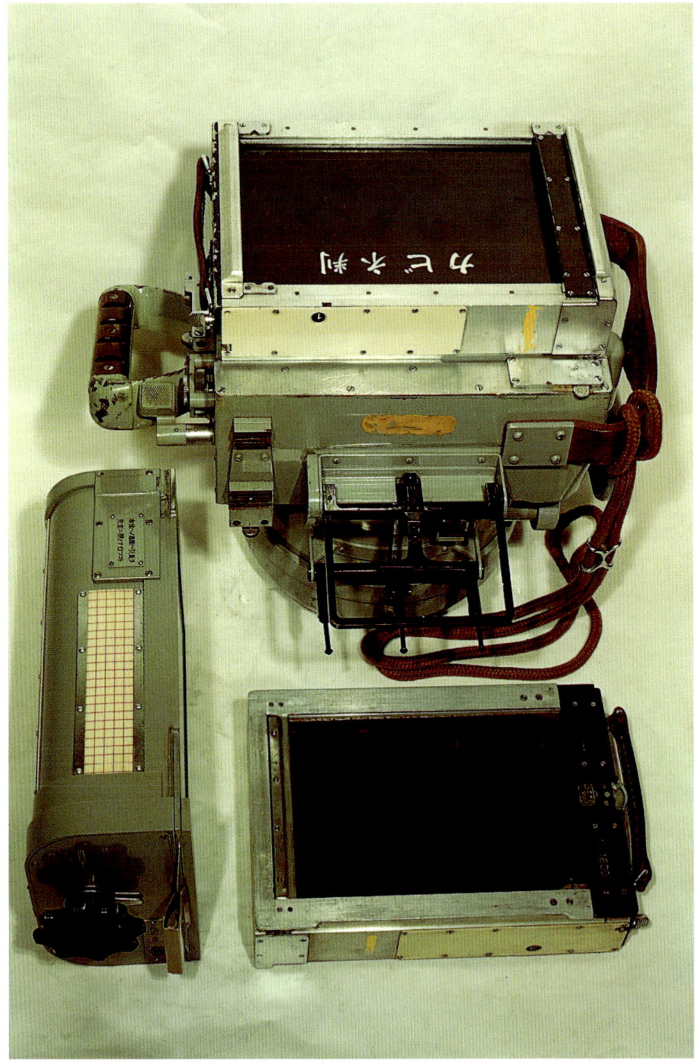

Army Type 96 Camera with plate film holder in place, two film holders, and roll film dispenser.

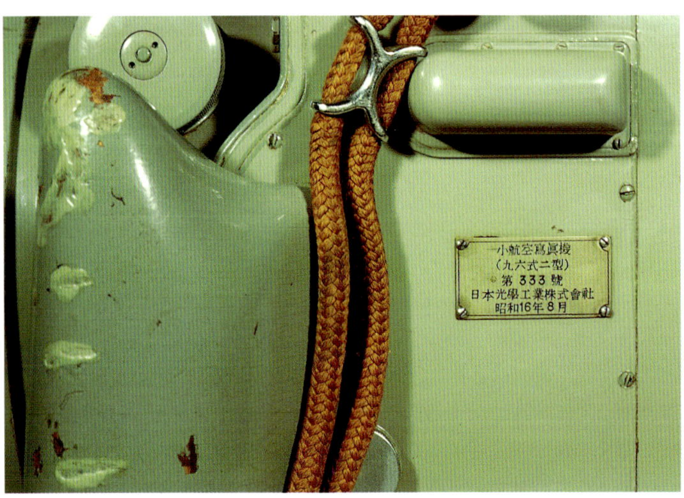

Army Type 96 Camera identifying placard.

Army Type 100 (SK-100) Small Aerial Camera used in Japanese Army Air Force reconnaissance aircraft.

Original carrying case for the Army Type 100 Small Aerial Camera showing nomenclature and specifications on box lid.

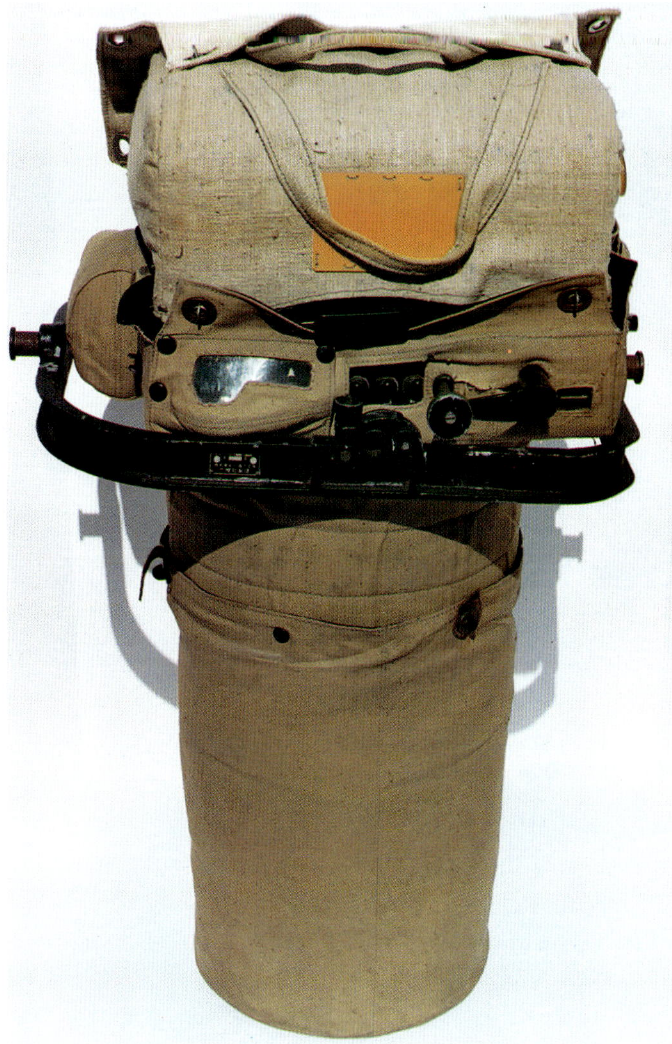

Army Mk.1 camera showing manual wind lever and on/off switches. Compare this to the similar Navy version of the K-8 camera.

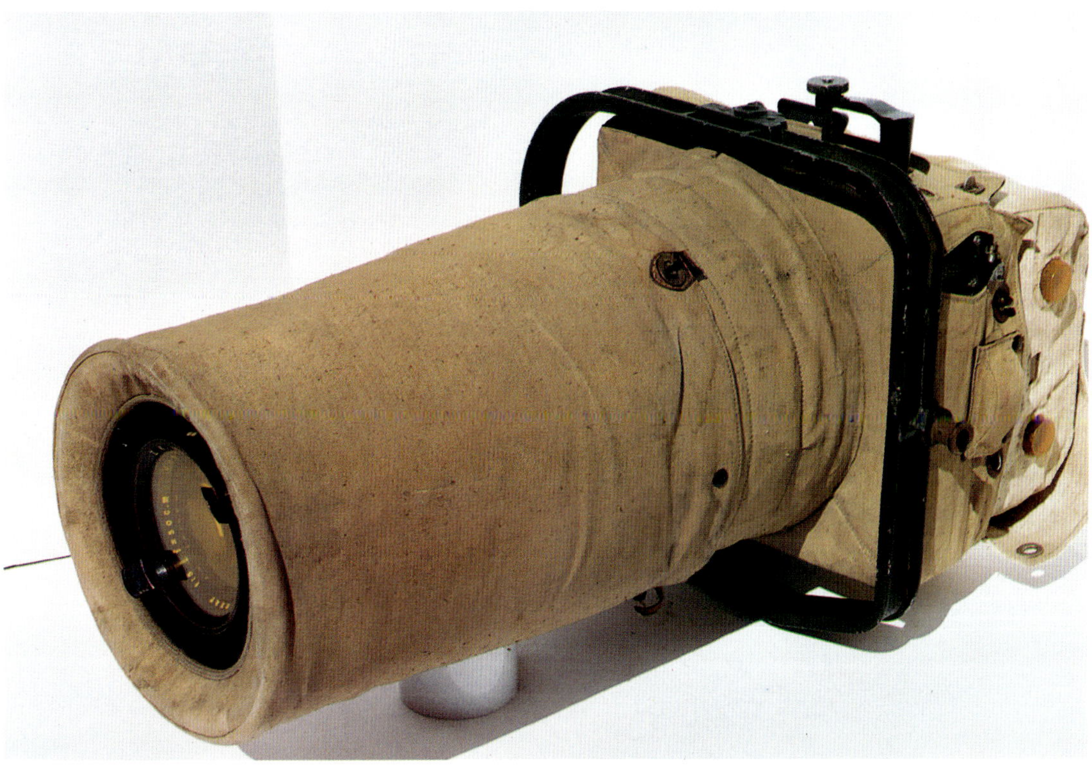

Japanese Army Mk.1 Camera with insulated jacket. Used for maping and strike photography.

photography. This Konica camera used roll film and came with several different lenses. Lens sizes included a 20 cm f3.5, 40 cm f5.6, and a 50 cm f5.6. The negative size is 12.5 x 16.5 cm, with 36 exposure capability. Its focal-plane shutter was marked for speeds of 1/200, 1/300, and 1/400 of a second. The camera's fitted carrying case holds four filters, a spare glass film plate, and a power cord for the camera's heating elements. The Type 100 was produced by Konica, *Chiyoda Kogaku* (later Minolta Camera), and *Nippon Kogaku* (Nikon). None of these cameras or accessories are marked with the date of manufacture. The Showa date block is present on the camera data plate, but not filled in.

The JAAF used the Mk.1 Automatic Aerial Camera for photo mapping and strike photography. This model was equipped with either a 25 or 50 cm lens, and also had insulation jackets to hold in heat during high altitude flights. The Mk.1 has many of the same specifications and design features as the Japanese Navy's K-8 camera. The Mk.1 used a 24-volt power supply vs. the 12-Volt supply for the K-8. The Army's Mk.1 camera could be used in manual operation or automatic operation using an electric motor. The camera had shutter speeds of 1/50, 1/100, and 1/150 of a second. The model illustrated has the 1.150 speed painted over in black, and this shutter speed may not have been available on this particular camera. The camera used 24 cm (9inch) roll film, taking images 16 cm x 24 cm. This example has a serial number of 1144 and a 50 cm f5 Konica Hexar Lens. An illustration depicts the lens with a bayonet mounted yellow filter marked "ZK. I D-52." There were several Army property marks (star in circle) on the cold weather covers. The Showa date is not present on the camera body's data plate. This example is battle damaged, with a 50-caliber bullet hole through the cold weather cover and lens cone. Konica was the main manufacture of the Mk.1, with a small number made by Nippon Kogaku (Nikon). The Nippon Kogaku examples were equipped with Nikkor 25 cm and 50 cm lenses.

This is a lower view of the Navy Type 99 Small Aerial camera. Note the serial number 1307 at the bottom.

Navy Cameras

The Type 99 was a hand-held oblique camera made for the Japanese Navy. This fixed focus camera was equipped with a 150 mm f4.5 lens. The camera used 70 mm roll film taking images 70 x 100 mm with 20 exposure capability. Its focal plane shutter was marked from 1/75 to 1/400 of a second shutter speed. It had a fitted carrying case with two lens filters (three were originally equipped) and a laminated instruction sheet. This model did not employ heating elements in the camera body or lens. Konica was the manufacturer.

The F-8 camera is similar to the Army Type 100 Small Aerial Camera. This Navy F-8 camera was, however, capable of using roll film or interchangeable plate film backs. All models examined were equipped with heating elements in the camera body. This camera

The Navy Type 99 Small Aerial Camera was hand-held, shown here with instruction booklet, lens cover, and two filters. Its paper contents list is dated August 1944.

was equipped with a 25 cm F4.5 lens. Konica manufactured the K-8 camera in several models. It is possible that the Japanese purchased rights to the American Fairchild F-8 and K-8 camera designs used by the USAAF and kept the designations, yet such a sale has not been confirmed (See page 113 showing the installation of these two cameras in the J1N1 reconnaissance Irving.).

The early models were equipped with a manufacturer's data plate, with the Showa date and model designation "F-8" on the camera body. This camera has several different features (weight and shutter speeds) than the late war version illustrated here. This later model has a laminated white data plate with the serial number and Konica's manufacture symbol (cherry blossom with figure inside). The later models do not include the date or the designation "F-8" on the camera body. However, the camera's carrying case includes a paper contents list designating the camera model as "F-8" and includes the Showa date. This model (serial number 2556) was made in November 1944.

The K-8 camera is very similar to the Japanese Army Mk.1 Automatic Aerial Camera. This camera was equipped with a 25

F-8 Aerial Camera used by the Japanese Navy showing *Konica's* manufacturing symbol and serial number 2556.

This F-8 Aerial Camera of the late war variety was manufactured by *Konica* for the Japanese Navy Air Force.

F-8 Aerial Camera, complete with power cord, filters, and plate back.

F-8 Aerial Camera showing plate film back. The design is that of the American Fairchild F-8 camera, thus the same designation was retained.

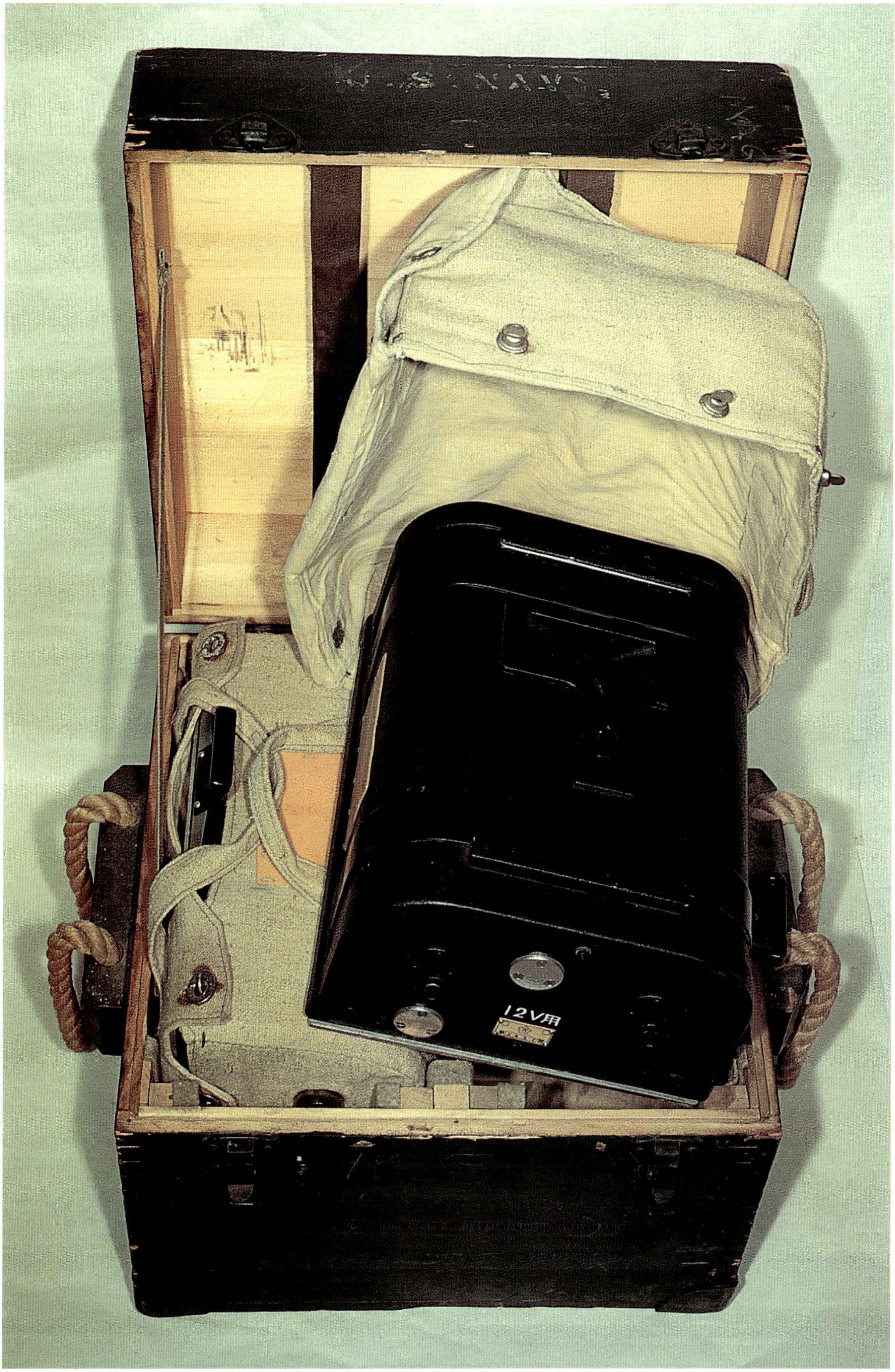

K-8 Aerial Camera in box manufactured by *Konica* for Japanese Navy Air Force.

K-8 Aerial Camera body showing drive on the side.

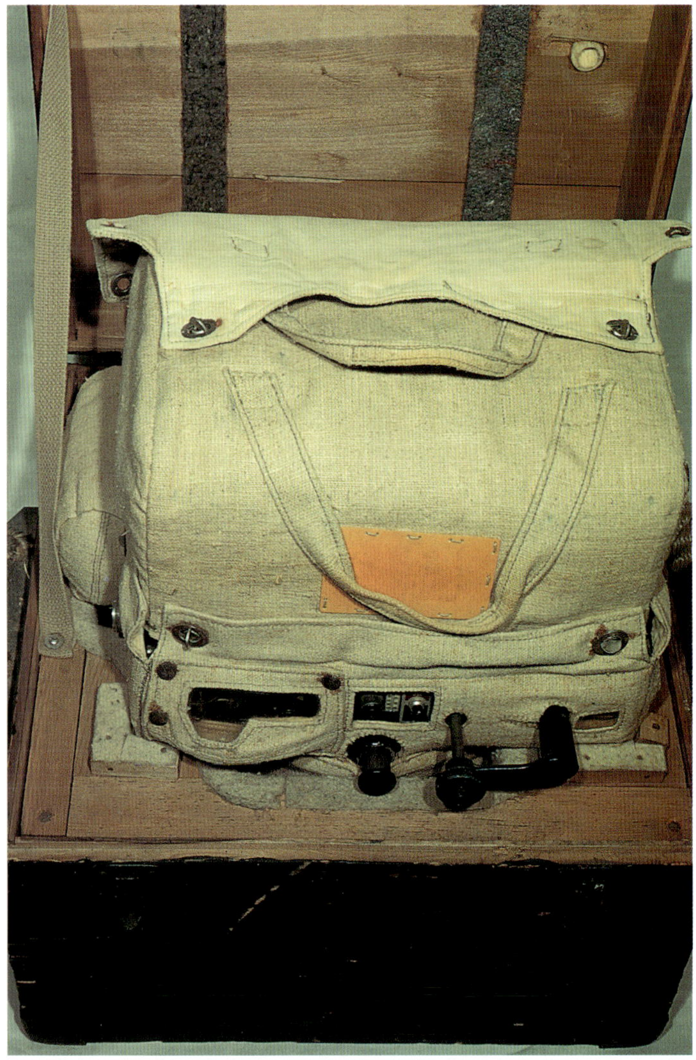

K-8 Aerial Camera with insulating jacket.

cm or 50 cm lens. The camera used 24 cm (9 inch) roll film taking images 18 cm x 24 cm. Like the Army Mk.1 Automatic Aerial Camera, the JNAF could use the K-8 in manual or automatic operation. These cameras were equipped with heating elements in the body, lens, and film back. The cameras were also equipped with insulating jackets designed to hold in heat during high altitude flights. The camera body has a film number counter that was projected onto the film plane when a photograph was taken. This example has a serial number of 987 and a 25 cm f4.5 Konica Perigon Lens. The Showa date is not present on the camera's data plate, but is included in the paper contents list in the box as June 1945. The K-8 camera mount (illustrated) enabled flexibility in positioning the camera during flight.

Close-up view of the K-8 Aerial Camera showing name plate.

The Japanese Navy made a series of experimental oblique cameras with various lenses. Additionally, they made a series of cameras designed for training purposes when used with practice torpedo bombing and dive bombing. The Japanese language document *General View of Aeronautical Engineering* states that these two cameras drew considerable interest by the U.S. Strategic Bombing Survey after the war. As a rule, Japanese aircraft did not carry aerial cameras unless the aircraft was specifically assigned for a reconnaissance mission.

Front view of the insulating jacket for the K-8 Aerial Camera showing the lens exposed.

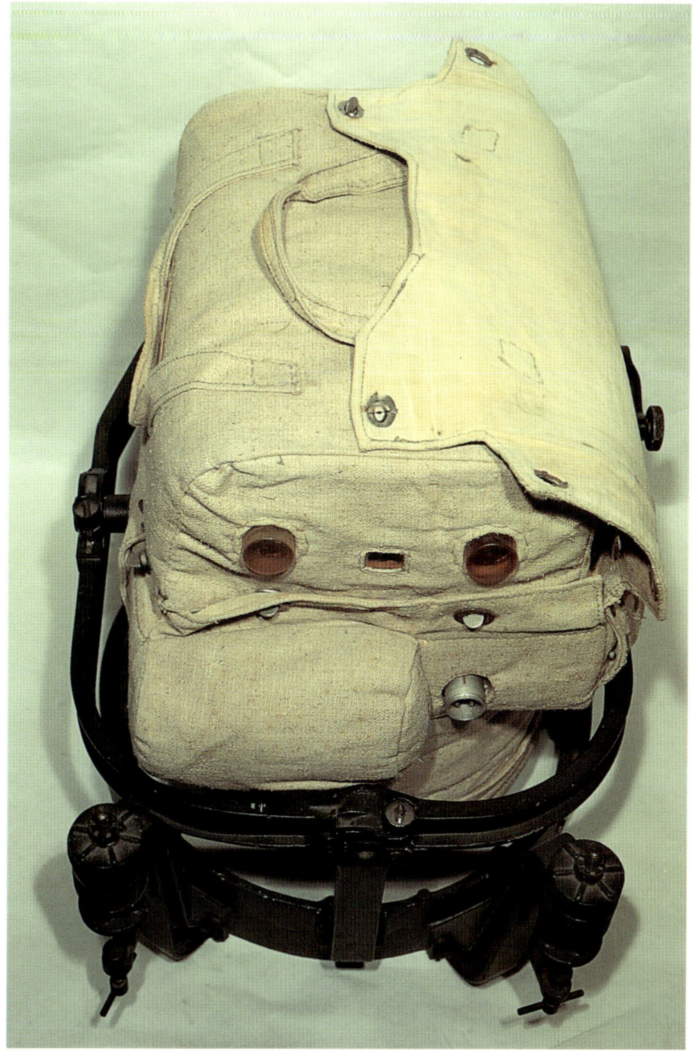

K-8 Aerial Camera in mounting rack and cold weather cover.

K-8 Aerial Camera body showing shutter speed dial and power switch.

K-8 Aerial Camera showing the mounting gimble, which supports the weight of the camera, yet allows freedom of movement.

This pre-war Japanese camera is illustrated for the record only. It contains no nomenclature other than "Handheld-type Aerial Camera," and therefore the branch of service is unknown. According to the placard, it was built by Rokuosha (now Konica) and has a 25 cm lense, with image size 13 x 18 cm. ***Courtesy of Keith Ashburn***

This Army gun camera resembles a machine gun, complete with bead and ring sight. Often referred to as a machine gun camera. It was used in lieu of live firing in flexible gun position locations.

Gun Cameras

Contrary to the previously mentioned intelligence report which states that the Japanese did not use fixed gun cameras operationally in combat aircraft, combat footage of this type was beginning to appear in Japanese wartime videos. Note, for example, the profile drawing of the Irving Escort Fighter that had a gun camera installed in the nose (p112). Pilots' personal observations, however, were reported as being the primary method for recording enemy plane losses. It is unclear if any flexible gun mounted cameras were used to record enemy engagements and losses.

A number of gun recording cameras were used for training purposes during the war. Literal translation for these cameras to record fixed gunnery was Firing Inspection Camera. One of these used by the JAAF was the Type 89, which resembled a flex-mounted machine gun, and several that resembled Lewis machine guns, as illustrated here. Two aircraft known to have used these cameras were Nate and Sonia.

This close up of the data plate for the machine gun camera identifies this as *Kaiten Shiki Syageki Kannsa* (Flexible Type Training [gun] Camera), Haisu Type (seemingly a foreign design or import name), s/n 1420, manufactured January 1943, by *Roku Ou Sya* (now *Konica*).

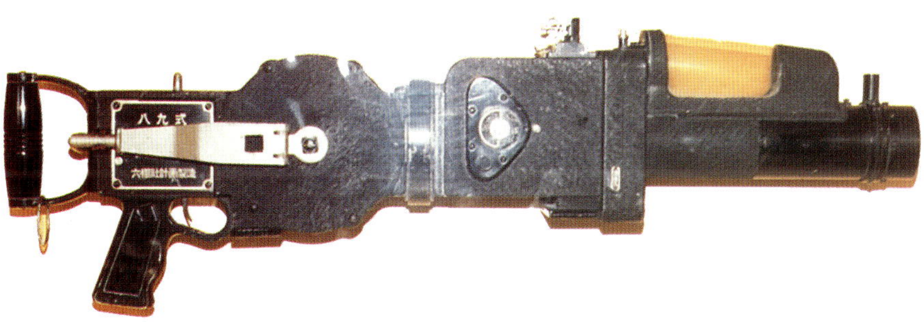

Another training aid was this Type 89 Movie Camera Gun of the Japanese Navy used for flexible gunnery training. This camera is on display at the Naval Memorial Museum in Washington, DC. *Courtesy of Todd Pederson*

This Army 35 mm Type 1 Fixed [gun] Training Camera is s/n 1548. This would record fighter aircraft gun firing. The date for manufacture has been left blank. The bottom line identifies the manufacturer as *Roku Ou Sya*, later to be *Konishi-Roku*, and known now as *Konica*.

Japanese Army Type 1 Gun Camera, 35 mm roll film cartridges, and power cords. One roll of film would record 50 images in simulated gun firing. ***Courtesy of Todd Pederson***

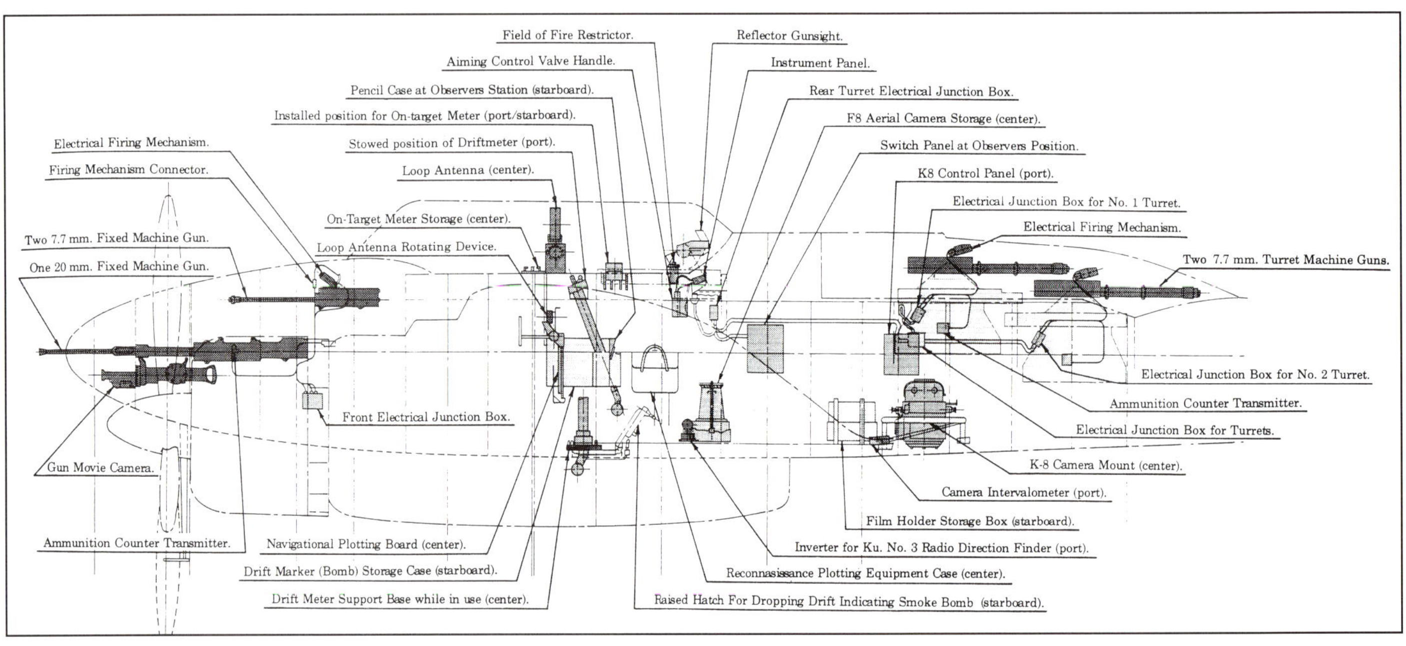

This inside profile of the Nakajima J1N1 Irving as a reconnaissance and escort fighter shows the installation of the nose gun camera and F-8 Cameras. Note also the armament when configured with twin turrets.

5

Machine Guns and Machine Cannon

by
Theodore E. Bradstreet

The application of the machine gun to military aircraft began for Japan with the siege of the German forces in China in late 1914. This operation also occasioned the first modification of a ground machine gun specifically for aircraft use. An early Army experimental light ground gun had its hand guards/bipod removed and was flexibly mounted to the cabane of a Nieuport NG2. With the efforts of the French and British air missions in Japan after WWI, both the Imperial Japanese Army and the Imperial Japanese Navy achieved an equal pace with the West in the use of aircraft machine guns. Most of the guns in use between the wars were imported, as were many of the aircraft themselves, but the Army developed its own flexible gun. Beginning in 1916, the Year 3 Type heavy infantry machine gun was extensively modified for flexible aircraft use. This gun was adopted in 1919 and served until replaced by the original Type 89 twin flexible (*see* below) in early 1930. The total separation between the army and naval services continued to apply with respect to aircraft armament throughout this period.

By the opening of the Pacific War, several guns formerly imported were license-built in modified form in Japan, original designs were in service as well, and progress had been made towards larger than rifle-caliber guns by both services, though most fighter kills were still made with cowl mounted, synchronized, rifle-caliber guns in WWI style. As the Pacific War progressed, the evolutionary trend was toward guns of larger and yet larger caliber to deal with both

Typical installation of the Type 92 flexible Lewis gun mounted in the nose of a Betty 11 Navy bomber. The ball mount used here was developed for tank machine guns to save both space and weight. ***80-G-192228***

Army Type 89 fixed: The cable that extends forward connected the gun to the synchronizer gear in the propeller transmission. The hardware attached at the rear of the receiver served to cock/charge the gun. Compare this image with that of the Navy Type 97 fixed gun below. ***80-G-193297***

the increased armor of opposing fighters and the growing size of Allied bombers. Fully automatic weapons of 57 mm caliber flew in combat aircraft by the end of the war.

During the Pacific War, as part of Western wartime racist propaganda, all Japanese machine guns were touted as inferior copies of Western designs, and this notion has tended to persist. Sometimes, in fact, the Western design allegedly copied often bore only a superficial resemblance to the Japanese gun. Most actual copying was licensed, with technical assistance from the foreign manufacturer, but there was less direct copying than commonly thought, combined with original design. Japanese-made guns often appeared ugly to Western eyes; standards of appearance differed. Certainly some Japanese-designed guns performed to less than Western standards, but those guns were generally not used in the same role as their Western counterparts. True copies of Western guns copied performance as well, even if they were not pretty. What mattered most was that the guns did what they were designed to do, however strange or rough their appearance, and they generally performed efficiently and effectively.

The following listings presuppose some knowledge of Pacific War Japanese research, design, and production organization, and of basic machine gun technology on the part of the reader. There is, however, a glossary after the listings, intended to be of assistance with terms that may be unfamiliar to some readers.

Service Guns by Caliber

Army Operational Guns

A note on nomenclature: The Army referred to all automatic firearms of 11 mm or less in caliber as *kikan juu*, "machine gun," and to those of greater than 11 mm as *kikan hou*, "machine cannon." That distinction is reflected in the designations used below. Also, about 1940 the Army instituted a system of designating aircraft machine gun design projects using the abbreviations *te* and *ho*. *Te*

This previously unpublished photograph shows the Army Te-1 (flexible version of the Type 89 fixed Vickers) installation in the tail of the Ki-21-Ib Sally. The tail cone and weather skirt have been removed, allowing a view of the cable controlled flexible mount and the feed and link ejection chutes. This mount allowed limited movement only. ***Courtesy of Shorzoe Abe***

Same Ki-21-Ib aircraft, but with tail cone and weather skirt in place. ***Courtesy of Shorzoe Abe***

is the kana abbreviation of *teppou*, "musket," or "military rifle," and was used for design projects of 11 mm or less in caliber ("Te-4" for example). *Ho* is the kana abbreviation for *hou*, and it was used for design projects of greater than 11 mm in caliber. Each gun was formally adopted and given a *shiki* (type) number; prior to this it was identified by its project number. Because the project number was unique to a specific gun and less cumbersome than the formal *shiki* designation, the project number often stayed with the gun in service.

Army Type 89 7.7 mm fixed machine gun

The original Type 89 fixed was a license-built copy of the Vickers Class E gun chambered for the then new 7.7 x 58 SR Type 89 cartridge (basically an improved .303" British). The Type 89 fixed gun synchronized well, but also saw much use as a wing mount. It was heavily revised in the 1938-40 period, and thereafter marked Type 89 *kai*. The Te-1 was a Type 89 fixed gun made flexible: it was used as a remote-control tail sting in the Ki-21, and perhaps in other applications.

In Service:	Type 89, 1930-*ca.* 1940; Type 89 *kai*, *ca.* 1940-1945; Te-1, 1939-1944 at least
Cartridge:	7.7 x 58 SR
Muzzle Velocity:	820 m/s
Action:	Short recoil, toggle locked (Vickers)
Feed:	Disintegrating belt
Rate of Fire:	900 rpm
Weight:	12.7 kg
Length:	1.035 m
Developer:	Vickers; *Giken* possibly
Known Maker:	Tokyo (early), Kokura and Nagoya Army Arsenals

Army Type 89 7.7 mm flexible machine gun

The Type 89 flexible machine gun (special)—the "Type 89 twin flexible gun" of the Pacific War—was derived from the original Type 89 flexible gun that entered service in early 1930. The original Type 89 twin gun was, in turn, derived from the *otsu-gou* flexible experimental of 1922-1929, a single gun very similar to the Te-4

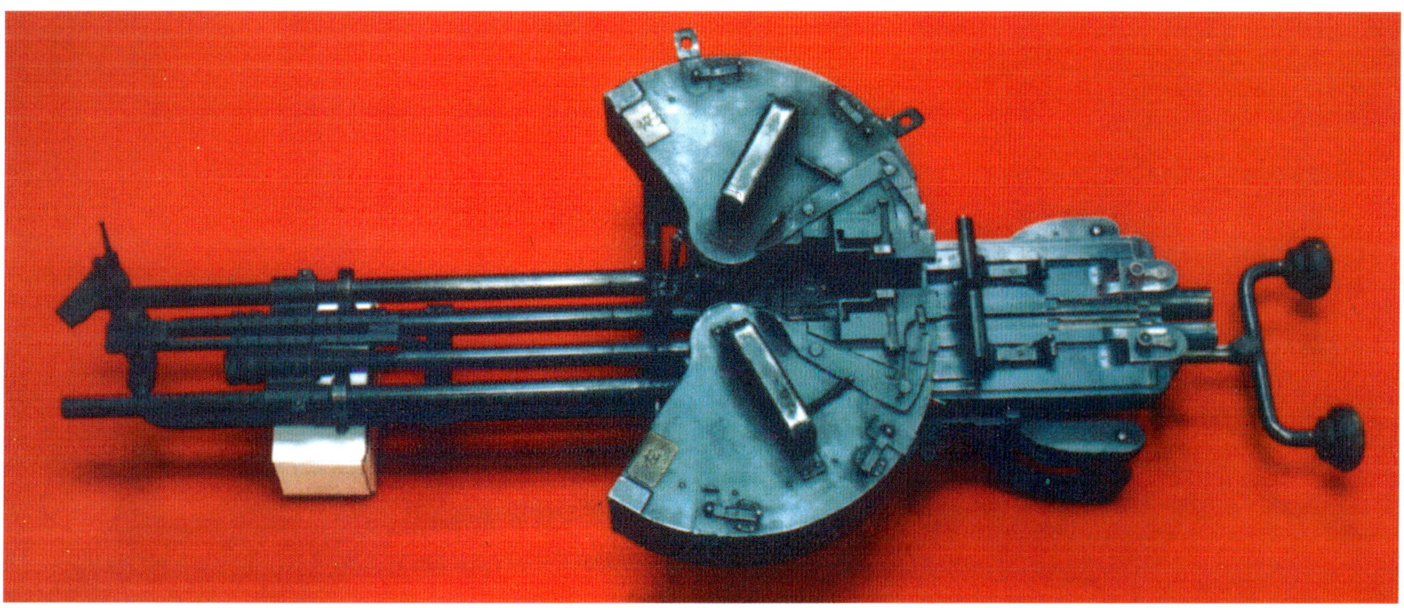

Army Type 89 flexible (special): Note the bulk and complexity of this gun in this top view. In spite of this, the high rate of fire of this twin gun made it attractive. Note also that the right and left guns are mirror images of one another. ***Courtesy of Edwin F. Libby***

This rare close in view shows the Army Type 89 (special) twin gun installed in the dorsal gun position of a Mitsubishi Ki-21-Ib Sally heavy bomber. Note the "vane" front sight. Note also the telescoping canopy that fully covered and streamlined the position in non-combat flight, and even provided the gunner some protection from the slipstream when open. ***Courtesy of Shorzoe Abe***

(seen here). The *otsu-gou* was essentially a Year 11 Type light infantry machine gun turned on its side and fed by a pan magazine. The Type 89 flexible gun therefore became two mirror image Year 11 Type actions fed by quadrant magazines and joined by a cradle carrying the controls (The Year 11 Type of designation was soon changed around this time to what became standard; Type 11, etc.). The original Type 89 flexible gun used ammunition on loose five-round clips in its magazines, and ejected the empty clips into the aircraft slipstream. This caused problems in nose mounts in multi-engine aircraft, so the Type 89 (special) retained the clips on a short captive belt inside each magazine.

In Service:	1930-1937? original Type 89; 1934-1945, Type 89 (special)
Cartridge:	7.7 x 58 SR
Muzzle Velocity:	810 m/s
Action:	Gas, falling block locked (Nambu)
Feed:	Unique detachable 90-rd quadrant magazine, (special) with internal belt
Rate of Fire:	1400 rpm
Weight:	28 kg
Length:	1.079 m
Developer:	Tokyo/possibly Kokura Army Arsenal
Known Maker:	Kokura and Nagoya Army Arsenals

Te-4

Despite its marking and its designation in official documents, this gun was confusingly referred to as "Type 89 flexible," often neglecting to mention the fact that it was a single, by the Japanese as well as the Allies. This makes the determination of flexible gun type used in some aircraft very difficult. Although the Te-4 never received a formal *shiki* designation, it was in limited production from before 1936 and served throughout the Pacific War. It is extremely similar to the *otsu-gou* experimental flexible gun from which the Type 89 twin flexible was derived, but it is marked "Type 89 (modified single)." This marking would seem to indicate its derivation by splitting a Type 89 twin, and such a split gun is known. Still, there are other indications that the Te-4 is directly derived from the *otsu-gou*. The Te-4 entered quantity production in 1941.

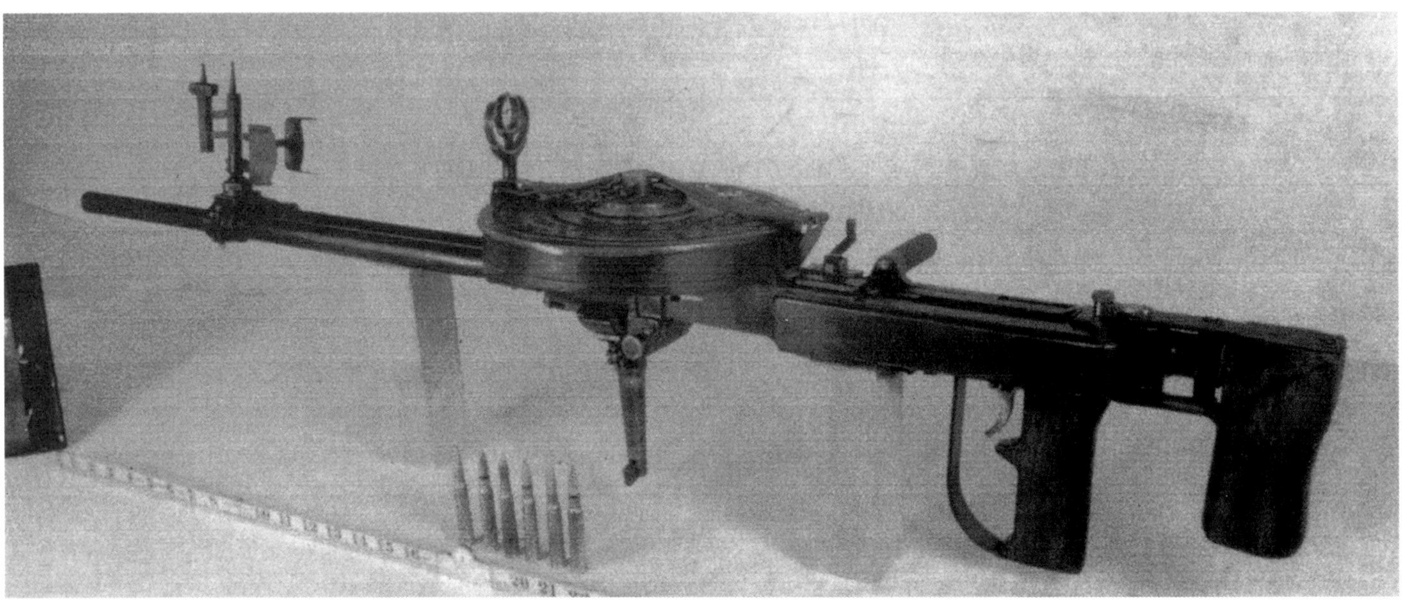

Army Te-4: The complex front sight, called a "vane" sight, was used from WWI on flexible guns in general. It is not often seen on Pacific War guns. It used the force of the slipstream to move the sight to compensate for aircraft speed in deflection shooting. Note that the slate refers to this gun as simply "Type 89 flexible machine gun," a common error. Empty cartridge cases were ejected downward into an accordion-pleated bag attached to the gun, not shown here. ***80-G-193293***

It appears that a GI has crawled into the trap door belly gun position of this Kawasaki Ki-48 Llly to man a slightly incomplete Te-4 for this photo. A Japanese crew member would likely have been a better fit, but not a particularly comfortable one. The Te-4 seems to have been standard in this particular application, although the Type 98 7.9 mm was used in other flexible gun positions in the same aircraft. The gun must have retracted in some way for the door to close.

In Service:	before 1936-1945
Cartridge:	7.7 x 58 SR
Muzzle Velocity:	810 m/s
Action:	Gas, falling block locked (Nambu)
Feed:	70-rd self-propelled pan magazine
Rate of Fire:	730 rpm
Weight:	9.3 kg
Length:	1.069 m
Developer:	Probably Tokyo Army Arsenal/*Giken*
Known Maker:	Tokyo (early), Kokura, and Nagoya Army Arsenals

Army Type 98 7.9 mm flexible machine gun

In 1938 the Army obtained licensing for and adopted the Luftwaffe MG 15 and MG 17 as the Type 98 flexible and Type 98 fixed aircraft machine guns, respectively. The Japanese could not produce springs for the guns that would sustain a long burst, however, so only the manually operated flexible gun (MG 15) was actually put into service. Apparently it was considered that it could be more readily restarted when it "ran out of gas." The Navy Type 1 flexible machine gun (see below) was essentially the same gun, and Nagoya Army Arsenal produced large quantities for the Japanese Navy late in the war. As Nagoya's production for the Army used the same tools and dies, late war Army guns are confusingly marked as Type 1, though they lack the Naval anchor mark.

In Service: 1938-1945
Cartridge: 7.92 x 57
Muzzle Velocity: 750 m/s
Action: Short recoil, rotating collar locked (Solothurn)
Feed: 75-rd saddle drum
Rate of Fire: 1100 rpm
Weight: 7.2 kg
Length: 1.078 m
Developer: Solothurn
Known Maker: Nagoya Army Arsenal

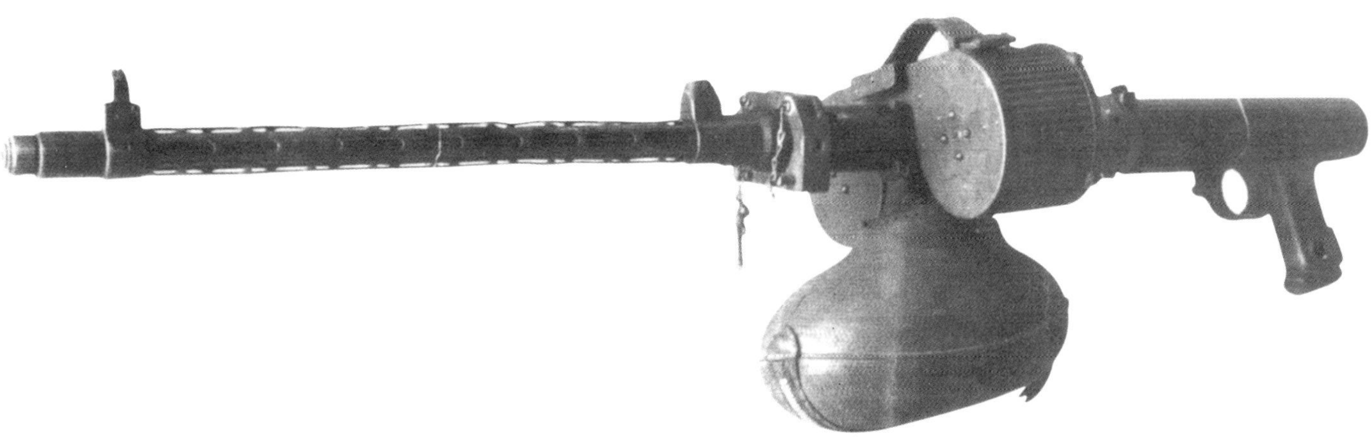

Army Type 98 flexible: Sights have been removed from this gun. Note the rugby-ball shaped empty case catcher attached to the gun below the magazine and closed by a zipper. ***80-G-193286***

The Nakajima Ki-49 Helen was more spacious, and thus better protected, than earlier Japanese Army bombers. Nevertheless, the cramped quarters of this Type 98 flexible gun position amidships shows why Japanese bombers in general had problems accommodating larger defensive weapons. ***80-G-169777***

Army Type 100/Type 1 7.9 mm flexible. This specimen is on a posing stand: the actual mounting point is near the front of the magazine under the receiver. The sight bases of this gun are empty. Note the lightweight chest brace found on some specimens of this gun. ***Courtesy of UK MoD Pattern Room***

Army Type 100/Type 1 7.9 mm flexible machine gun

This gun (a twin gun) was developed by placing the actions of two Czech ZB 26 (BREN type) infantry guns side by side in a single receiver—a remarkable piece of work. There is no difference between the Army Type 100 and Army Type 1 gun whatsoever, rumors to the contrary, although some guns did have a detachable chest brace. The distinction is simply a nomenclature change; when the Ho-103 was adopted as the Type 1 fixed (see below), the Type 100 flexible was redesignated to make a Type 1 fixed/flexible pair, parallel to the Type 89 and Type 98 pairs. Virtually all Type 100 guns and accessories, such as toolboxes, were re-marked to Type 1 by having the zeros individually struck out by diagonal lines.

In Service:	1940-1945
Cartridge:	7.92 x 57
Muzzle Velocity:	750 m/s
Action:	Gas, tilting bolt locked (ZB 26)
Feed:	Unique 100-rd saddle drum, each 50-rd side feeding only that side of the gun
Rate of Fire:	2200 rpm
Weight:	16.7 kg
Length:	1.18 m
Developer:	*Giken*
Known Maker:	Nagoya Army Arsenal

Army Type 1 13 mm fixed machine cannon (Ho-103)

During the search for a medium caliber aircraft gun in 1938-39, Italian guns in 12.7 x 81 SR were tested. The guns were not adopted, but the Army liked the cartridge as made by the Italians. The Ho-103 was a project to adapt the U.S. Model 1921 aircraft Browning gun to the Italian cartridge—no mean feat. The starting point was indeed the Model 1921 and not the M2 Browning; this is clear from detailed parts comparisons. The gun was scaled down and a number of changes made, the most important change being the addition of a

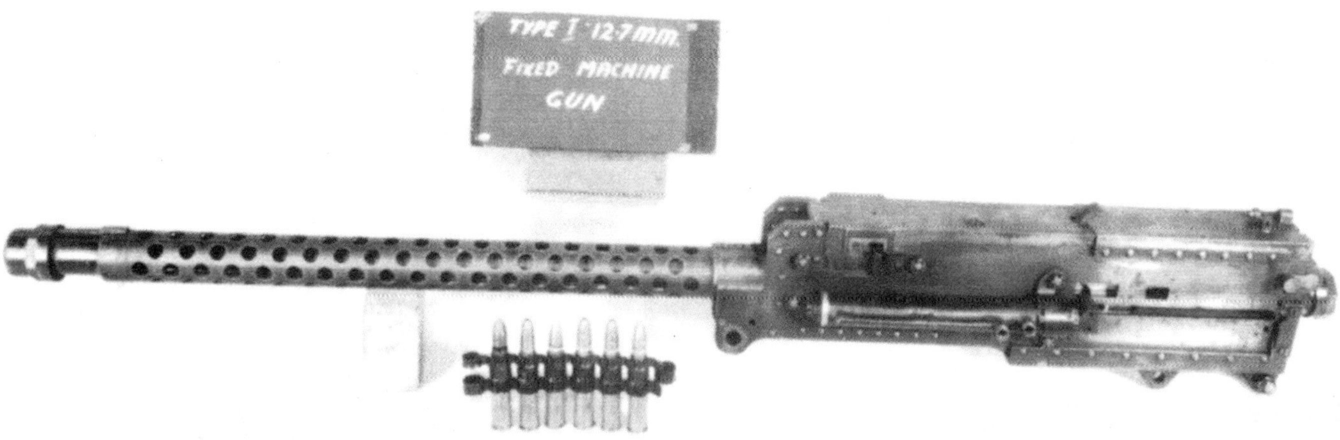

Army Type 1 fixed (Ho-103): Note the perforated barrel jacket supporting the moving barrel to the muzzle, typical of most Browning guns. The muzzle attachment beyond the jacket—the muzzle booster—is not typical, however. ***80-G-193291***

This pair of synchronized Ho-103s in a Kawasaki Ki-61 Tony cowling are mounted with a slight stagger, the port gun slightly ahead of the starboard. This arrangement is most likely necessitated by ammunition storage and feed requirements in the tight and busy space of the Ki-61 nose. *80-G-169744* and *80-G-169745*

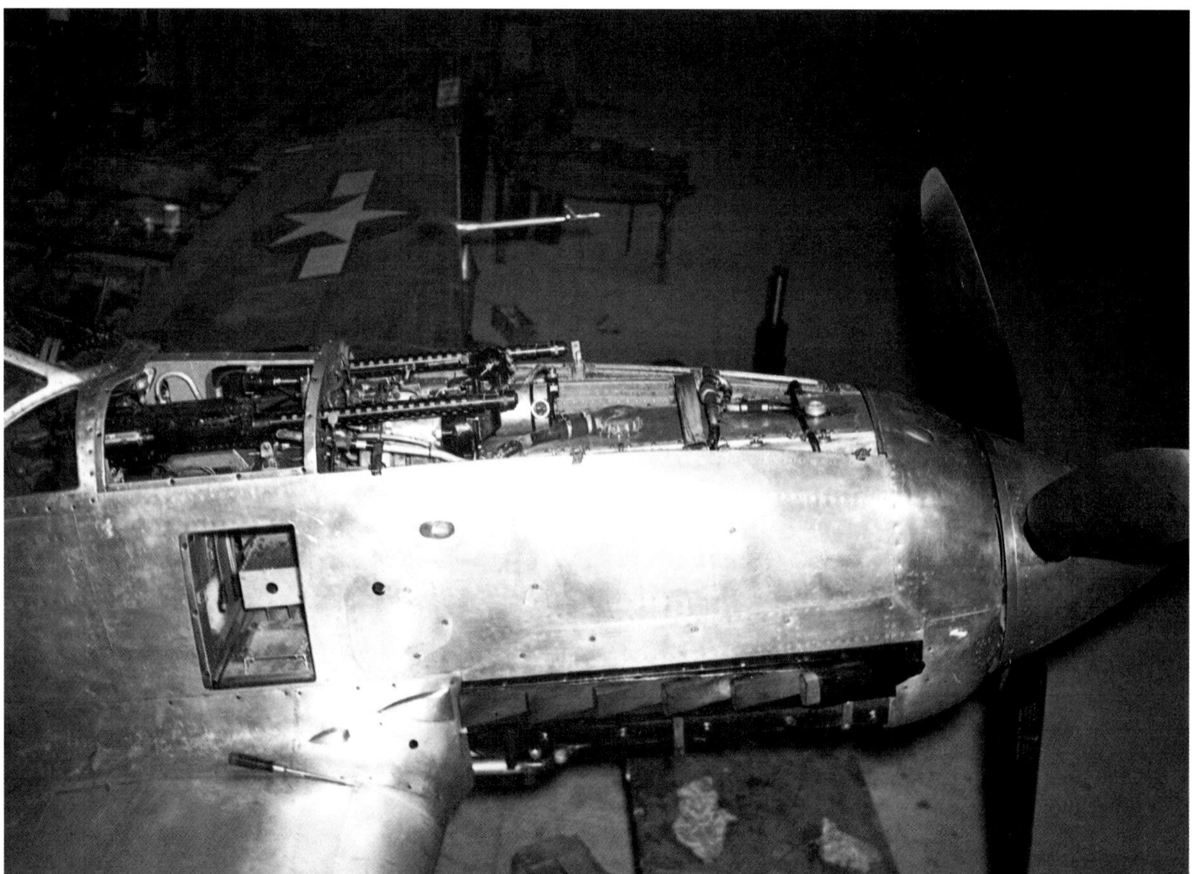

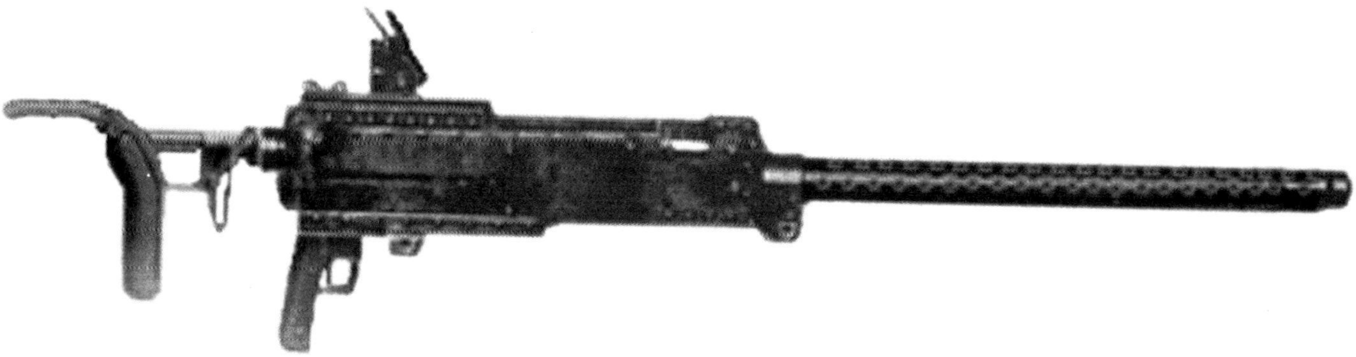

Army Type 1 13 mm flexible (Ho-104): This gun was derived from the Ho-103 but was distinct from it. Note the different receiver shape, particularly at the upper front corner, and the different application of the muzzle booster. This specimen has a Type 3 reflector gunsight mounted on it. *Funatsu Aviation Indicator Museum*

muzzle booster to allow the much less powerful cartridge to operate the gun. The result was very successful but, like all Brownings, the gun did not synchronize well, losing much of its rate of fire. The fixed gun was used as a flexible by mounting it in a cradle, to which all manual controls were attached. A true flexible version, the Ho-104, was developed late in the war and saw service in the Ki-67 bomber 1944-45, but length and weight data for it are unavailable at present. Performance is given as the same as the Ho- 103.

In Service:	1941-1945
Cartridge:	12.7 x 81 SR
Muzzle Velocity:	780 m/s
Action:	Short recoil, rising block locked (Browning)
Feed:	Disintegrating belt
Rate of Fire:	900 rpm unsynchronized, as low as 400 rpm synchronized
Weight:	23 kg
Length:	1.267 m
Developer:	Chuuou Kougyou KK
Known Maker:	Chuuou Kougyou KK; Kokura and Nagoya Army Arsenals

Ho-1

This was the first in a series of Ho-00X 20 mm aircraft machine gun projects. The Ho-1, like its sister the Ho-3, was based on the powerful Army Type 97 anti-tank (AT) gun, and it is often also referred to as "Type 97." However, the Type 97 AT was semi-automatic only, and had a seven-round box magazine, while the Ho-1 was full-auto with a fifteen-round double drum. The actual Type 97 was never mounted in an aircraft. The Ho-1 even dispensed with the recoil cradle of the Type 97, as the Ho-1 was mounted in a complex mechanical—though open—turret with its own recoil management system.

In Service:	*ca.* 1941-1944 at least
Cartridge:	20 x 125
Muzzle Velocity:	820 m/s
Action:	Gas, rising block locked (Nambu)
Feed:	15-rd Kawamura double drum
Rate of Fire:	400 rpm
Weight:	30 kg
Length:	1.742 m
Developer:	Japan Special Steel/*Giken*
Known Maker:	Kokura Army Arsenal

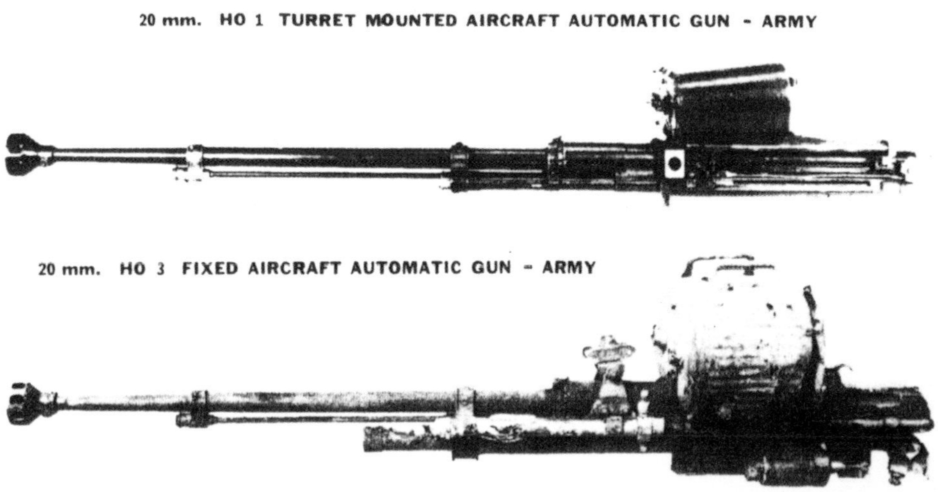

Army Ho-1 and Ho-3: The similarities between these two guns due to common derivation from the Type 97 antitank rifle are obvious in the barrels and their attachments, but the differences in magazines are also clear. Also, the large recoil buffer retained from its Type 97 ancestor is obvious under the barrel of the Ho-3, and just as obviously lacking in the Ho-1. *TAIC Manual No. 1, July 1945 supplement*

Ho-3

Also derived from the Army Type 97 AT gun, this weapon is likewise referred to as "Type 97," although neither it, nor its sister, the Ho-1, were ever given a *shiki* number. Unlike the Ho-1, the Ho-3 retained the recoil cradle of the Type 97 AT. The Ho-3 also used a much larger fifty round double drum magazine not interchangeable with the Ho-1 magazine. The slow rate of fire of these guns precluded effective use in fighter-fighter combat, but the Ho-3 was of some use as a standoff weapon against bombers because of its power and range. Still, it was often removed in service to save weight.

In Service:	*ca.* 1941-1945
Cartridge:	20 x 125
Muzzle Velocity:	820 m/s
Action:	Gas, rising block locked (Nambu)
Feed:	50-rd Kawamura double drum
Rate of Fire:	400 rpm
Weight:	45 kg
Length:	1.74 m
Developer:	Japan Special Steel/*Giken*
Known Maker:	Kokura Army Arsenal

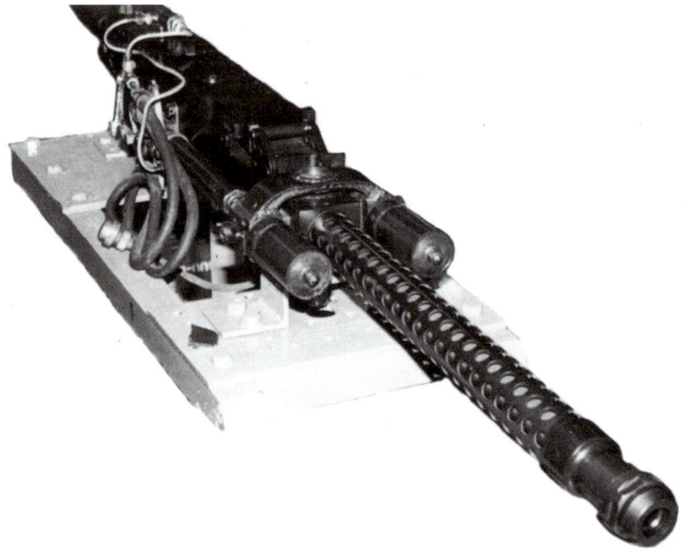

This display of a non-synchronized Ho-5 in the UK Ministry of Defense Pattern Room shows many details of a typical fixed mount of a larger Japanese Navy Browning. The twin cylinders beside the barrel are recoil shock absorbers. The longer, slimmer cylinder along the starboard side of the gun is the hydraulic charger, fed by the two hoses. The two fine tubes going to the device on the top of the gun provide hydraulic power for the trigger. ***Courtesy of Flying Guns: World War II, A. G. Williams and E. Gustin***

Army Type 2 20 mm lightweight machine cannon (Ho-5)

This was the first true fighter-fighter 20 mm cannon produced by the Army, the best Japanese 20 mm aircraft gun, and arguably the best WWII 20mm produced by any combatant nation—when made with decent raw materials and skilled labor. Unfortunately, both the latter ran swiftly short after the introduction of this weapon, necessitating the progressive downloading of its ammunition propellant charge to avoid parts breakage. This gun evolved by scaling up the Ho-103 as little as possible and chambering it for a shortened version of the very popular 20 x 110 Hispano-Suiza cartridge. The intent was to produce a 20 mm gun that would fit where the Ho-103 had fit, and it was nearly achieved. The Ho-5 was made flexible by mounting it in a cradle the same way as the Ho-103, but it seems to have been used little in this manner. Both true flexible and lightened versions of this gun were attempted without apparent success. Like other Brownings it did not synchronize well, and cowl mounts were often removed in service.

In Service:	Possibly 1942 -1945
Cartridge:	20 x 94
Muzzle Velocity:	820 m/s (early) - 750 m/s (late)
Action:	Short recoil, rising block locked (Browning)
Feed:	Disintegrating belt
Rate of Fire:	850 rpm (early) - 750 rpm (late, both unsynchronized)
Weight:	36.8 kg
Length:	1.40 m
Developer:	Chuuou Kougyou KK
Known Maker:	Hitachi Heiki KK; Kokura and Nagoya Army Arsenals

Army Type 2 lightweight fixed machine cannon (Ho-5): The derivation of this weapon from the Ho-103 (shown previously) is obvious, despite the less-than-clear quality of this image from a March 1945 supplement to ***TAIC Manual Number 1***

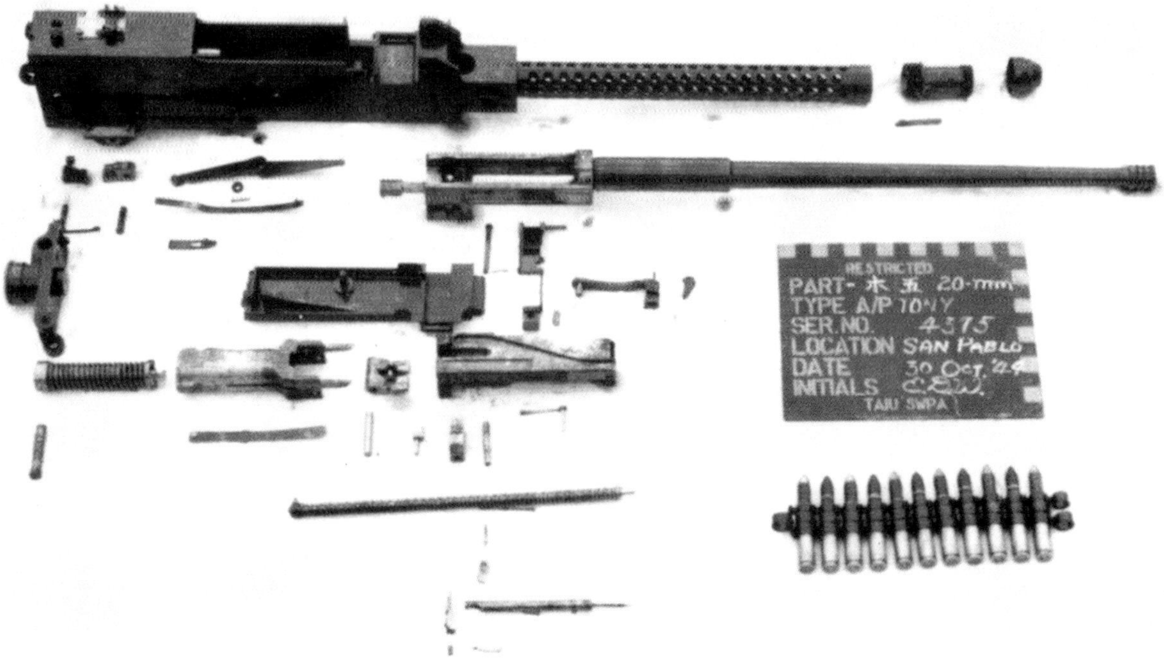

This Army Type 2 machine cannon (Ho-5) was photographed by a TAIC team showing the weapon and its various parts. This particular gun was obtained from a Ki-61 Tony for both nose and wing guns. *80-G-191360*

Ho-155

The Ho-155 II *gata* ("model," as actually marked on the gun) presently appears to be the only Army 30 mm that actually may have seen combat during the Pacific War, although it was co-developed and co-produced with the somewhat larger Ho-155-I. The Ho-155 prototype was a modification of the experimental-only 25 mm Ho-51, differing considerably from the "production" guns. While the Ho-155-I was scaled up from the Ho- 51, the Ho-155-II was trimmed down from the I *gata* in an effort to make the large gun fit in spots originally intended for the Ho-5; the receiver was reshaped and the barrel shortened. Many plans were laid for the Ho-155, but few can be confirmed as accomplished; in particular, it would have been very difficult to synchronize at a useful rate of fire. Because they entered the arena so late in the Pacific War, both Ho-155 guns were plagued with reliability problems, as well. Neither Ho-155 was ever formally adopted, both remaining in experimental status. These guns have been given the misnomer "Ho-105," among others, by some writers. Figures given below in brackets [ ] are for the Ho-155-II where different.

In Service:	very late 1944-1945
Cartridge:	30 x 115
Muzzle Velocity:	700 m/s
Action:	Short recoil, rising block locked (Browning)
Feed:	Disintegrating belt
Rate of Fire:	500 rpm (unsynchronized)
Weight:	50 kg [44 kg]
Length:	1.93 m [1.51 m]
Developer:	Chuuou Kougyou KK
Known Maker:	Nagoya Army Arsenal

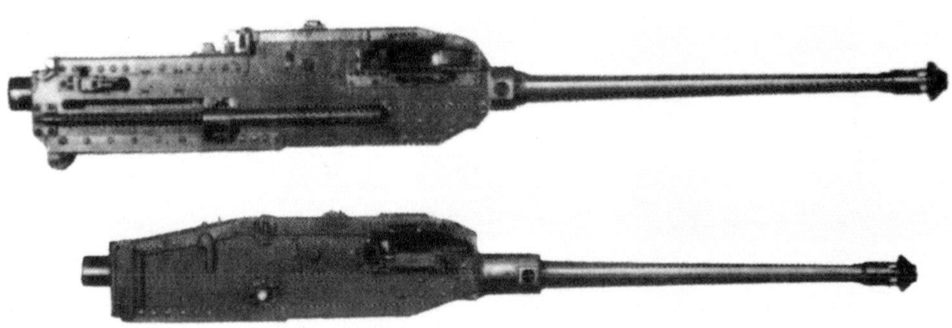

Army Ho-155: The Ho-155-I (upper) and Ho-155-II (lower) are shown here approximately to scale. Note the considerable reshaping of the Ho-155-II receiver and the lack of barrel jacket and distinctive muzzle brake of both. These latter two characteristics make Ho-155 installations in aircraft easy to recognize. *Images from G. M. Chinn, The Machine Gun*

Army Ho-203: The large "birdcage" magazine on the top of this gun is a non-removable part of the feed system, and also the reason why wing mounting of this gun was not practicable. Note the considerable amount of barrel exposed ahead of the barrel jacket. This feature serves to distinguish the Ho-203 from the Ho-401 in photos where size is not obvious. ***US Army Ordnance Intelligence, U.S. National Archives***

Ho-203 (perhaps Type 3 37 mm machine cannon)

It is not altogether clear whether the Ho-203 was actually given a *shiki* designation, as a few original Japanese sources refer to it as 3 *gata* rather than 3 *shiki*. It is essentially an automated version of the Year 11 Type direct-fire infantry gun and, as such, it is necessarily quite bulky. At some 18 inches (0.46 m) thick at the midriff, it was entirely unsuited for wing mounting, and proposals for the same never came to fruition, despite claims made elsewhere. Its very low rate of fire made it relatively useless in fighter-fighter conflict, though it was pressed into such use. It was, however, designed and installed specifically for bomber attack, at which it had continued success.

In Service:	1943-1945
Cartridge:	37 x 111 R
Muzzle Velocity:	570 m/s
Action:	Long recoil, artillery-style falling-block breech
Feed:	15-rd removable belt in non-detachable "birdcage" magazine
Rate of Fire:	Variable by changing buffer fluid, 130 rpm typical
Weight:	89 kg
Length:	1.52 m
Developer:	Japan Special Steel
Known Maker:	Japan Special Steel; Kokura and Nagoya Army Arsenals

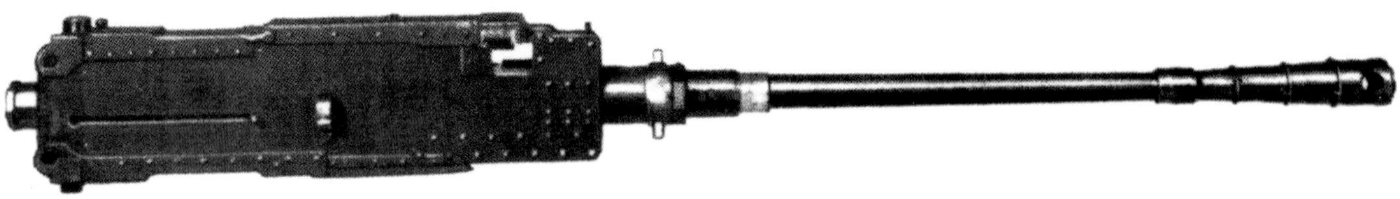

Army Ho-204: Note the forward extension of the rear barrel bearing on this largest of short-recoil Brownings. Note also the lack of barrel jacket and the distinctive ice cream cone shaped muzzle brake. The reinforcing rings around the muzzle brake are not present on all specimens. Muzzle brakes on the larger Brownings (Ho-155 and Ho-204) indicate that recoil forces generated by the cartridges used were more than necessary to operate the gun, in contrast to the smaller Brownings (Ho-103 and Ho-5), which needed muzzle boosters. ***Image from G. M. Chinn, The Machine Gun***

Ho-204 (perhaps Type 4 37 mm machine cannon?)

The uncertainty as to nomenclature here is the same as with the Ho-203. The Ho-204 was the largest short-recoil Browning ever to see combat service. Its size limited its application, and its apparent lack of combat success follows on that application. Primary use of this gun was in the Ki-46-III *kai* air defense fighter, an aircraft type that was unsuccessful because its poor rate of climb made it unable to actively intercept Allied bomber formations (Some sources misidentify the oblique gun in the Ki-46-III *kai* as the Ho-203, not a practical application of that gun.). The Ho-204 utilized a layout for Browning-based guns original to Japanese design. Rather than a barrel jacket providing support at the muzzle for the barrel, it and other large Japanese Brownings (*ie* the Ho- 155) extended the front of the receiver somewhat, lengthening the rear barrel bearing for barrel stability and eliminating the jacket to save considerable weight.

In Service:	1944-1945
Cartridge:	37 x 144
Muzzle Velocity:	450 m/s
Action:	Short recoil, rising block locked (Browning)
Feed:	Disintegrating belt
Rate of Fire:	400 rpm
Weight:	130 kg
Length:	2.47 m
Developer:	Chuuou Kougyou KK
Known Maker:	Chuuou Kougyou KK (few); Kokura and Nagoya Army Arsenals

Ho-301

Though not a spectacularly successful type, the Ho-301 is notable as the only aircraft gun using caseless ammunition to see combat to date. Elimination of the cartridge case allows for a much simpler and more compact gun design, since no empties must be extracted or ejected. This first effort with such a cartridge was of very low power, however—little more than a scaled-down automatic version of the Army Type 89 heavy grenade launcher ("knee mortar")—and it should really be considered an automatic grenade launcher. Very low muzzle velocity and resulting highly curved trajectory made this a very close range weapon. Low rate of fire combined with trajectory severely limited usefulness in fighter-fighter work, and forced very close approaches to bomber targets.

In Service:	1943-1945
Cartridge:	40 x 129 caseless
Muzzle Velocity:	245 m/s
Action:	Blowback with advanced primer ignition (Oerlikon)
Feed:	10-rd detachable box magazine, not removed in loading
Rate of Fire:	475 rpm
Weight:	40 kg
Length:	1.39 m
Developer:	Chuuou Kougyou KK
Known Maker:	Nagoya Army Arsenal

Army Ho-301: This specimen is in uncocked, or fired, condition. In cocked condition, all of the exterior parts of the gun forward of the receiver would be drawn rearward, compressing the action spring seen surrounding the barrel. This would give the gun a radically different appearance, exposing a considerable length of barrel, only the tip of which is here exposed. This specimen is mounted on a test plate fabricated at Aberdeen Proving Ground. ***Image from G. M. Chinn, The Machine Gun***

Army Ho-401: Compare this gun with the Ho-203 on page 125. Though it is a much larger gun, the similarities are obvious. Note the very short length of barrel beyond the barrel jacket, however, and the muzzle brake. This muzzle brake is of the same design as, though much larger diameter than, that of the Ho-155. Photo is inverted to operating position. *U.S. Army Technical Intelligence, U.S. National Archives*

Ho 401

The Ho-401 appears here because of a single B-29 kill claim by a ferry pilot of a brand new Ki-102b. While the claim is very doubtful, to say the least, it helps to put this gun into the service category, and its aircraft was in service, even if actually withheld from combat. The Ho-401 can be thought of as either a scaled-up Ho-203 or an automated Type 90 tank gun. It differs from the Ho-203 most obviously in size, and the addition of a mushroom shaped muzzle brake to counteract excess recoil.

In Service:	possibly 1944 -1945
Cartridge:	57 x 121 R
Muzzle Velocity:	560 m/s
Action:	Long recoil, artillery-style falling-block breech
Feed:	17-rd captive belt in non-detachable birdcage magazine
Rate of Fire:	variable by changing buffer fluid, 80-100 rpm
Weight:	160 kg
Length:	2.043 m
Developer:	Japan Special Steel
Known Maker:	Japan Special Steel; Nagoya Army Arsenal

Navy Operational Guns

Notes on nomenclature: The Navy used only one term, *kijuu*, "machine gun," to refer to automatic firearms of any size used in aircraft. Not only is the Japanese term distinct from Army terminology, but it also implies that even weapons of 40 mm caliber were small arms by Naval standards. The Navy did not use design project numbers that were specific to particular designs. Rather, the Navy issued general requirements for development programs under which several different guns might be developed. The programs were given *shi* numbers, and the guns were identified by the program number and other details, such as caliber and designer. *Shi* is an abbreviation for *shisaku*, "experimental," and was prefixed with the last two digits of the Showa year to make up the program designation. *Shi* designations did not generally follow guns into service. The Navy used *gata*, "model," and *kai*, an abbreviation of *kaizo*, "modification," as hierarchical modifiers of *shiki* to indicate variants: 97 *shiki* 3 *gata kai* 2 should be read as "Type 97 Model 3 Second Modification," and understood as "the second modification of the third model of the Type 97."

Navy Type 92 7.7 mm flexible machine gun

The Type 92 flexible was a license-built copy of the aircraft version of the Lewis light machine gun as built by BSA in England and previously imported. Once under production in Japan it did undergo some small modification; the version listed in an official Navy table dated January 1945 was only *kai* 1, so little change was made during the long career of the weapon. Of course, the gun was usually finished to standard Navy specifications; a salt spray resistant combination of a tough gloss black lacquer over a greenish-gray zinc phosphating, rather than the British high blue.

Navy Type 92 flexible: The enlarged trigger guard is one of the few Japanese modifications of the original Lewis design. The sights present here are very simple, typical of late flexible gunsights after the complex and expensive vane sights had passed from use. *80-G-193294*

In Service:	1932-1945
Cartridge:	7.7 x 56 R (.303 British)
Muzzle Velocity:	740 m/s
Action:	Gas, rotating bolt head locked (Lewis)
Feed:	97-rd Lewis "drum"
Rate of Fire:	600 rpm
Weight:	8.5 kg
Length:	0.98 m
Developer:	Savage Arms Co.
Known Maker:	Toyokawa and Yokosuka Naval Arsenals

Navy Type 97 7.7 mm fixed machine gun

The Type 97 fixed gun was a license-built copy of the Vickers E class aircraft machine gun. It went through rather more modification in Japanese hands than the Lewis, as the current version in January 1945 was 3 *gata kai* 2, and there may subsequently have been a *kai* 3. The gun synchronized very well, was a favored cowl mount, and was responsible for many kills by Zeros.

In Service:	1937-1945
Cartridge:	7.7 x 56 R (.303 British)
Muzzle Velocity:	745 m/s
Action:	Short recoil, toggle locked (Vickers)
Feed:	Disintegrating belt
Rate of Fire:	900 rpm
Weight:	12.6 kg
Length:	1.033 m
Developer:	Vickers
Known Maker:	KK Nihon Seikoujo; Suzuka and Yokosuka Naval Arsenals

Navy Type 1 7.9 mm flexible machine gun

Like the Army Type 98 flexible machine gun, this gun was a license-built copy of the Luftwaffe MG 15. In early production, the Navy Type 1 differed from the Army Type 98 in having the typical Naval lacquer-over-phosphate finish and carrying a second handgrip below the forward end of the receiver. However, in a perhaps unique

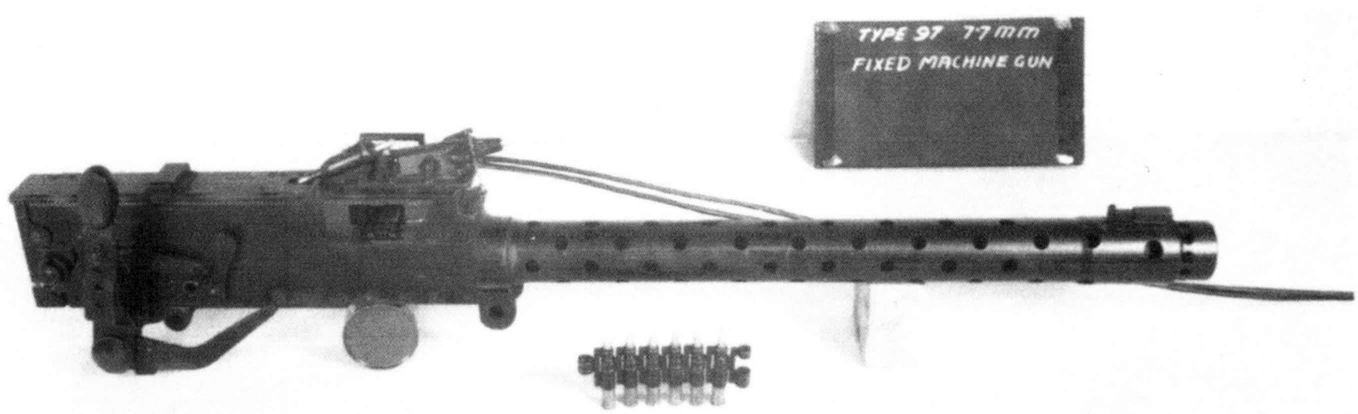

Navy Type 97 fixed: Very similar to the Army Type 89 fixed gun (compare above) because of shared ancestry. The Navy Type 97 differed mainly in its cartridge, continuing to use the original .303" British. The cable leading forward connected with the synchronizing gear, and the lever at the rear served to cock/charge the weapon. *80-G-193302*

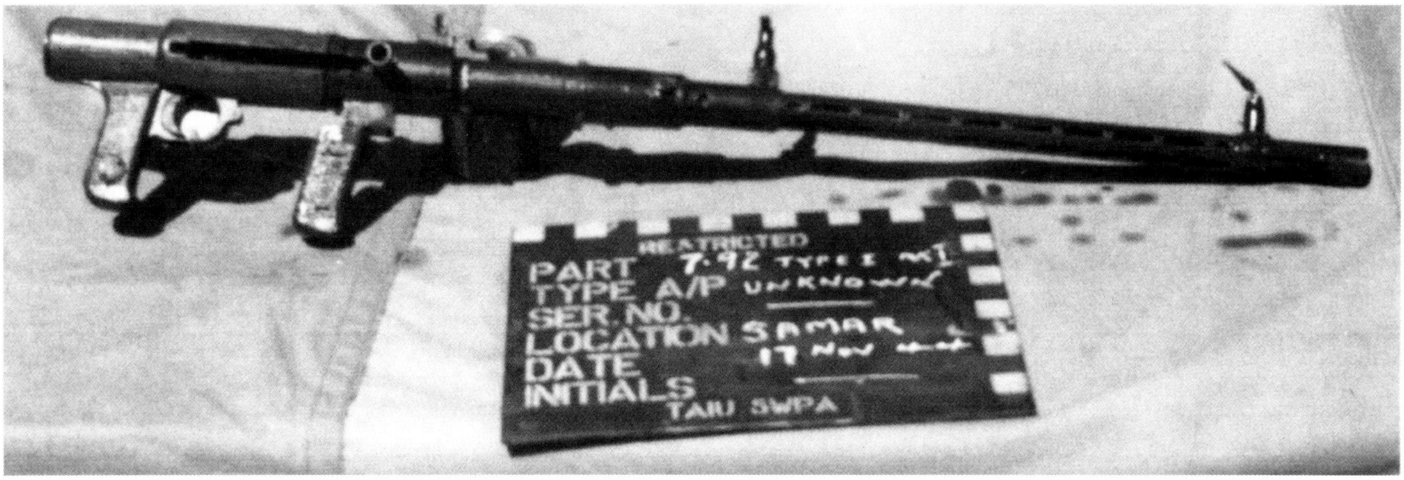

Navy Type 1 flexible: Although the sights are damaged and the magazine is missing, this image serves to show an important point of difference between the Navy Type 1 flexible and the Army Type 98 flexible, both MG 15 copies. That difference is the auxiliary handgrip under the receiver, present only on earlier, Navy-built specimens. ***80-G-191275***

instance of Army/Navy cooperation, Nagoya Army Arsenal became a major manufacturer of the Navy Type 1. Perhaps starting before this point, and certainly afterward, the distinction broke down. Finish became simply zinc phosphate, the second handgrip disappeared, and both Army and Navy guns made by Nagoya were marked Type 1, the only distinction being the Naval anchor mark. Aside from such changes, and some minor alterations to simplify production, no changes were made to the original Solothurn design.

In Service: 1941-1945
Cartridge: 7.92 x 57
Muzzle Velocity: 785 m/s
Action: Short recoil, rotating collar locked (Solothurn)
Feed: 75-rd saddle drum
Rate of Fire: 1000 rpm
Weight: 6.9 kg
Length: 1.077 m
Developer: Solothurn
Known Maker: Nagoya Army Arsenal; Tagajou and Yokosuka Naval Arsenals

Navy Type 2 13 mm flexible machine gun

The Navy Type 2 flexible machine gun was a license-built copy of the Luftwaffe MG 131 in the original percussion-ignition form of that gun. The Navy did not feel that electric ignition could be expected to be reliable in the tropical marine environment, where rapid corrosion of critical contacts was the rule, so purchased the earlier version. The Navy contract for the gun with Rheinmetall-Borsig included technical assistance not only with the gun, but with several specific projectile types, as well. No significant changes were made to the gun in Japanese service, though there was some substitution of wood for plastic parts. The gun was produced as a free gun only, probably because of the same difficulty with spring manufacture that prevented Army production of the Army Type 98 fixed (MG 17) machine gun.

In Service: possibly1942-1945
Cartridge: 13 x 64 B
Muzzle Velocity: 750 m/s
Action: Short recoil, rotating collar locked (telescoped Solothurn)
Feed: Disintegrating belt

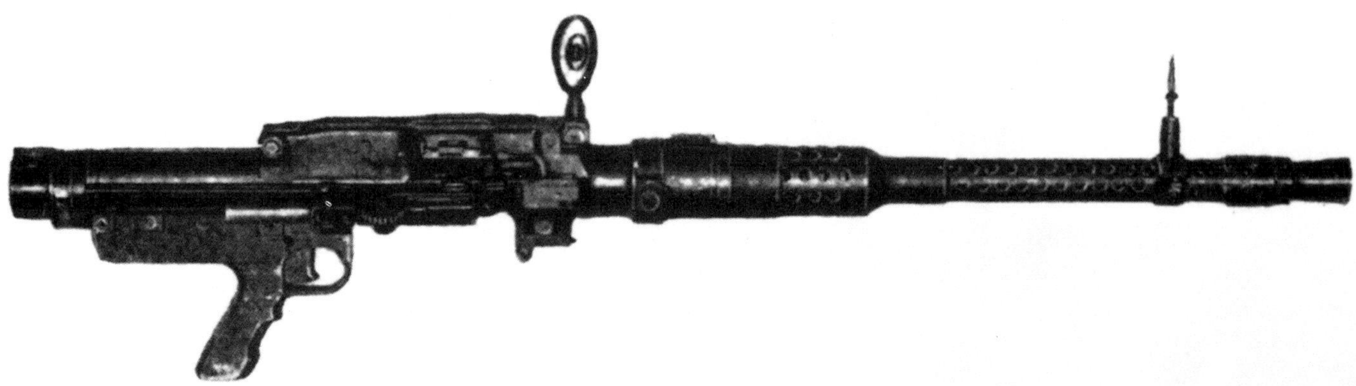

Navy Type 2 13 mm flexible: Little distinguishes this gun from its German counterpart externally. The simple aluminum slab grip and IJN style sights are the only giveaways here. ***U.S. Army Signal Corps, U.S. National Archives***

Navy Type 3 fixed: The cylinder with a wire hanging from it at the rear of the receiver is the firing solenoid; electrical firing was typical of many remotely fired guns. The conical device at the muzzle is often called a flash hider. It does not actually hide the flash, but reshapes it from a large, blinding, round ball into a narrower form, easier on the pilot/gunner's eyes at night. *80-G-191470*

Rate of Fire:	900 rpm
Weight:	17.4 kg
Length:	1.17 m
Developer:	Rheinmetall-Borsig AG
Known Maker:	KK Nihon Seikoujo; Suzuka (possibly) and Toyokawa Naval Arsenals

Navy Type 3 13 mm fixed and flexible machine guns

The Navy Type 3 guns were essentially straightforward copies of the M2 Browning in contemporary U.S. service, but chambered for the 13.2 x 99 Hotchkiss cartridge then in Navy service, rather than for the very similar 12.7 x 99 Browning. In fact, .50 BMG ammunition—belt and all—could be, and evidently was, fired in Type 3 guns *in extremis*. Figures in brackets [ ] below are for the flexible gun where different.

In Service:	1943-1945
Cartridge:	13.2 x 99
Muzzle Velocity:	790 m/s [800 m/s]
Action:	Short recoil, rising block locked (Browning)
Feed:	Disintegrating belt
Rate of Fire:	800 rpm
Weight:	27.5 kg [34 kg]
Length:	1.53 m [1.55 m]
Developer:	*Kugishisyo* possibly
Known Maker:	KK Nihon Seikoujo; Suzuka and Toyokawa naval Arsenals

Navy Type 99 Mk.1 20 mm fixed and flexible machine guns

All the variants of the Navy Type 99 Mk.1 gun developed in its long service were based on a license-built copy of the Oerlikon FF. In 1935, when the three sizes of FF-based Oerlikons—FF, FFL, and FFS—were brand new, farsighted Navy-connected individuals very quietly brought all three into Japan. By 1937 this group had formed Dainihon Heiki KK, with the assistance of Oerlikon and support of some very highly placed Naval officials, and licensing was obtained

Navy Type 99 Mk.1 fixed model 4: This image shows a number of interesting points. First, the gun is in a cocked condition, with the reciprocating outer parts withdrawn well backward on the barrel compared to the Navy Type 99 Mk.1 flexible model 11 that follows. Second, the firing solenoid (cylinder with wire attached) and pneumatic charger (long cylinder lying alongside the receiver) are present. Third, the Kawamura belt feed is in place atop the gun. *80-G-191263*

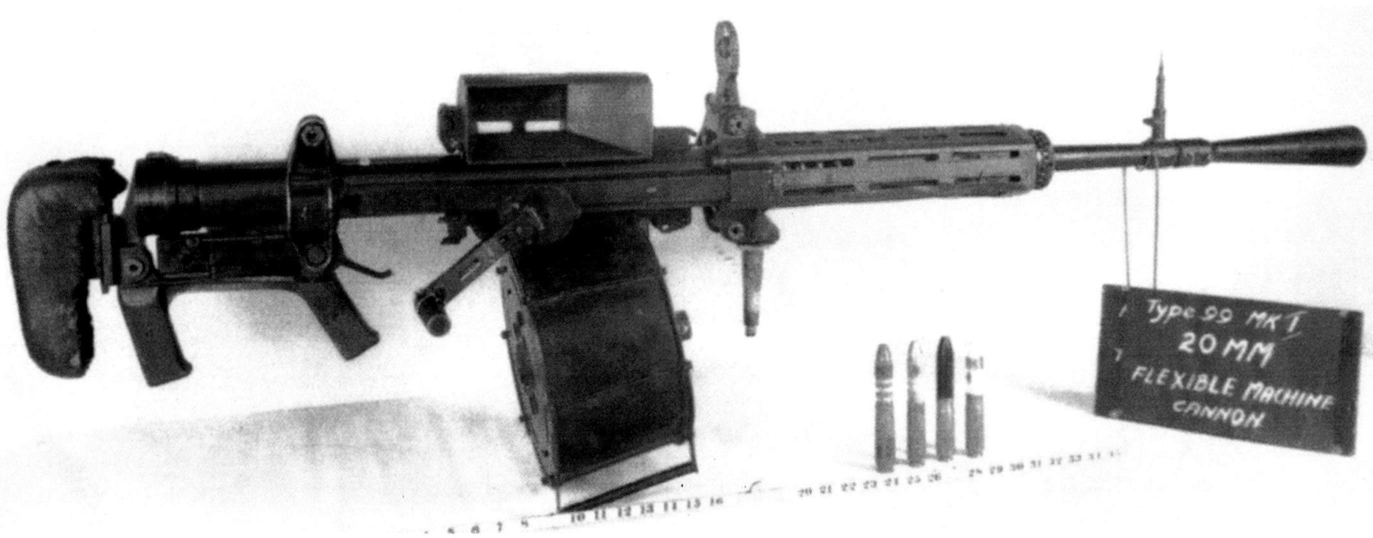

Navy Type 99 Mk.1 flexible model 11: Basically a Type 99 Mk.1 fixed gun with added accessories, this is the freely mounted version of the Type 99 Mk.1. Note the added handgrip/shoulder rest/trigger mechanism, hand-ratcheted charger, and forward handguard over the moving parts. Note also that the gun is "upside-down;" with the magazine on the bottom and ejection occurring at the top through a deflector. The free gun also has a particularly long, narrow flash hider to protect the gunner's vision at night. ***80-G-193290***

for all three sizes of Oerlikon FF guns. By 1939 Dainihon Heiki had full-blown manufacturing capacity at its Tomioka works. Nomenclature for the Oerlikon guns went through three phases. Earliest for the FF was *E shiki* 1 *gata* (old style; *E* for *Erikon*, the Japanese rendering of Oerlikon), followed in turn by *E shiki* 1 *gou* (new style, pre-adoption) and, lastly, on formal adoption in 1941, 99 *shiki* 1 *gou*, translated here as Type 99-1, but often seen as Type 99 Mark 1. The relatively low muzzle velocity of the Type 99-1 resulted in a curved trajectory that made hits difficult and the gun unpopular with pilots, at least at first. The Type 99-1 fixed gun went through four models (*gata*): the flexible; two models with two modifications (*kai*) of the last model of free (all-manual) guns; and at least two flexible models specialized for mechanical turrets. The data below is therefore given in ranges.

In Service:	1939-1945
Cartridge:	20 x 72 RB
Muzzle Velocity:	600 m/s
Action:	Blowback with advanced primer ignition (Oerlikon)
Feed:	30, 45, 60, and 100-rd drum; Kawamura disintegrating belt feed (Model 4 only)
Rate of Fire:	520-550 rpm (Model 4 fixed belt-fed fastest)
Weight:	23.2-30 kg (free gun heaviest because of accessories)
Length:	1.331-1.653 m (free gun longest because of flash hider, *etc.*)
Developer:	Oerlikon/Dainihon Heiki KK
Known Maker:	Dainihon Heiki KK; Toyokawa Naval Arsenal; likely others

Navy Type 99 Mk.2 20 mm fixed machine gun

The history of the Navy Type 99 Mk.2 fixed gun closely parallels that of the Type 99 Mk.1, since the Type 99-2 was a license-built Oerlikon FFL, the intermediate size of Oerlikon FF-based guns. The Type 99-2 entered service somewhat later, as airframes became roomier and stronger to accommodate its increased size and recoil. Early designations that apply to it are *E shiki* 2 *gata* (old style) and *E shiki* 2 *gou* (new style). There were no free-gun versions of the Type 99-2, and all flexible applications used fixed guns in mechanical turrets. There were, however, five models of the fixed gun. The Model 5 was only briefly in service (adopted in May 1945), and may not have seen combat, but is particularly interesting. It used the Kawamura belt feed of the Model 4 and a very strong multiple coil spring external bolt buffer to achieve a much higher rate of fire than typical for Oerlikon guns. Its specifications are in brackets [] below.

In Service:	1941-1945
Cartridge:	20 x 101 RB
Muzzle Velocity:	750 m/s [760 m/s]
Action:	Blowback with advanced primer ignition (Oerlikon)
Feed:	Possibly 30 and 45, definitely 60 and 100-rd drum; Kawamura disintegrating belt feed
Rate of Fire:	480-500 rpm [620 rpm]
Weight:	33.5-37.57 kg [38.5 kg]
Length:	1.89 m [1.855 m]
Developer:	Oerlikon/Dainihon Heiki KK [*Kugishisyo* possibly]
Known Maker:	Dainihon Heiki KK; Tagajou Naval Arsenal

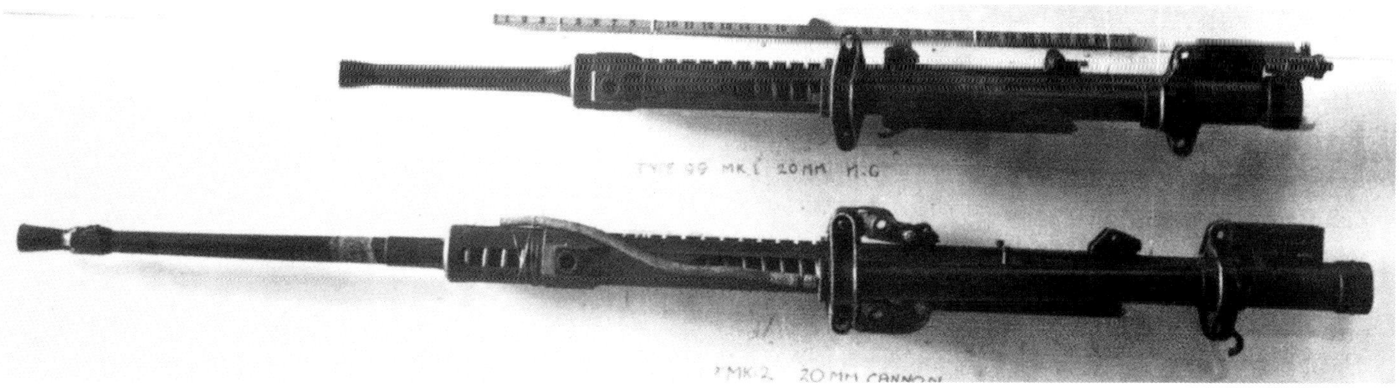

Navy Type 99 Mk.1 and Type 99 Mk.2 fixed models 3: This image is provided to show relative size of the Type 99 Mk.1 (upper) and 99 Mk.2 (lower) fixed guns. Both appear to be Model 3 of their respective types, i.e. magazine-fed. ***80-G-193284***

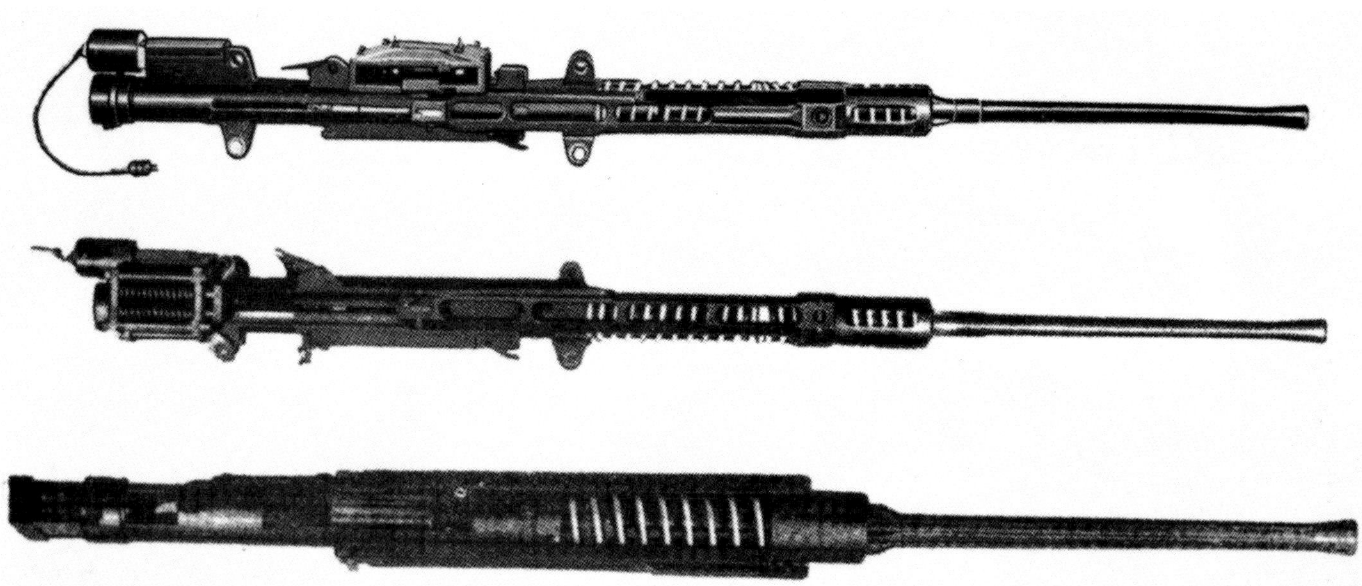

Navy Type 99 Mk.2: The upper gun is a Type 99 Mk.2 Model 4 with the Kawamura belt feed in place. The center gun is the Type 99 Mk.2 Model 5 without its belt feed, and the bottom gun is the 30 mm Type 2 Oerlikon by way of comparison. Note the large coil-spring bolt buffers added to the Model 5, the shortening of its receiver with respect to the trigger solenoid, and the amount of action spring visible through additional lightening openings in the action sidebars. Note also how difficult it might be to distinguish an installed Type 2 from a Type 99 Mk.2. The barrel of this Type 2 specimen is fluted, but not all may have been. ***Images from G. M. Chinn, The Machine Gun***

An angular perspective of the Navy Type 99 Mk.2 Model 4 machine cannon is best for showing many identifying features. This is the type known as the 20 mm wing cannon for the Zero fighter.

A number of oblique-mounted gun installations were made to various aircraft for use as night fighters. Here is a D4Y2 Judy photographed at Atsugi AB having a Navy Type 99 Mk.2 Model 4 cannon in the rear compartment. The housing at the sides is the feed and link-ejection chutes for the Kawamura ammunition belt feed. *Courtesy of James P. Gallagher*

The open bay door provided access to the two upper Type 99 Mk.2 Model 3 guns of this Nakajima J1N1-S Irving night fighter for loading the (absent) 100-round drum magazines. Note the staggering of the guns and the port gun mounted slightly ahead of and less deeply in the fuselage than the starboard, to allow room for both magazines.

Type 2 30 mm fixed machine gun

The Type 2 30 mm fixed machine gun was somewhat of an embarrassment to the *Kohon*, adopted more because of the political influence of Dainihon Heiki KK within the Navy than out of any real desire for the gun. Underpowered, slow firing, and of limited magazine capacity, it saw very limited deployment. The gun came into being because Dainihon Heiki was made aware of the 17-*shi* program (see under *Notes on Experimental Guns* below) early in 1942 before the *Kohon* had even issued its general requirements to the *Kugisyo*, and well before the *Kugisyo* had issued specific requirements for the program. Dainihon Heiki sought a time advantage over other designers by simply scaling up the basic Oerlikon FF gun to 30 mm. Because they did not have the requirements of the 17-*shi* program in hand, however, the result lacked the required power, rate of fire, and belt feed. Production was slow to start, with only a handful of guns produced in 1943. Combat testing, originally scheduled for Rabaul in August 1943, did not actually take place until spring 1944 on Truk, delaying final adoption. Even then, the gun was rejected for fighter wing mounting, ostensibly because of its small magazine capacity.

In Service:	1944-45
Cartridge:	30 x 91 RB
Muzzle Velocity:	710 m/s
Action:	Blowback with advanced primer ignition (Oerlikon)
Feed:	42-rd drum
Rate of Fire:	400 rpm
Weight:	50.9 kg
Length:	2.068 m
Developer:	Dainihon Heiki KK
Known Maker:	Dainihon Heiki KK

Type 5 30 mm fixed machine gun

On April 13, 1945, the 17-*shi* 30 mm fixed machine gun Model 1 (b) was chosen as the final product of the 17-*shi* program (see in *Notes on Experimental Guns* below), and adopted as the Type 5 30 mm fixed machine gun Model 1. This gun truly can be said to be wholly of original Japanese design, and was singled out by the U.S. Strategic Bombing Survey as "original and efficient." It achieved the goals of the 17-*shi* program (possibly excepting rate of fire) and

Navy Type 2 30 mm fixed: This is the same image as shown above as the bottom weapon, but perhaps the barrel fluting can be seen more clearly. Unfortunately this is a top (or bottom, it can be hard to differentiate with an Oerlikon gun) view of the gun without its large drum magazine. *Image from G. M. Chinn, The Machine Gun*

Navy Type 5 fixed: The forward mounting ears have been cut off this specimen for reasons unknown. The top plate, which carries the obvious feed linkage, supported the entire gun. The gun proper, barrel, receiver, and action parts internal to the receiver oscillated about 30 mm backward and forward on the top plate with each shot fired to achieve floating firing. ***Courtesy of Kenneth A. Huddle***

more. By firing each shot while the gun proper was still moving forward in counter-recoil from the previous shot ("floating firing"), the recoil force passed to the airframe was greatly reduced. Efforts were made to utilize this gun in whatever fashion in whatever aircraft type could accept it, and it was the weapon of choice for all Navy proposed and prototype fighter/interceptor aircraft of 1945. About 2000 guns were made, but it seems few got into action before the end of the war.

In Service:	1945
Cartridge:	30 x 122
Muzzle Velocity:	760 m/s
Action:	Gas-unlocked blowback with floating firing, Kjellman locked
Feed:	Disintegrating belt
Rate of Fire:	500 rpm (nominal)
Weight:	66 kg
Length:	2.092 m
Developer:	Japan Special Steel
Known Maker:	Toyokawa Naval Arsenal; test quantities by Japan Special Steel and KK Nihon Seikoujo

A computer-enhanced close-up of the installation of the Type 5 in the *Kyokkou* above. Note what appear to be the remains of a rubberized fabric boot at the junction of the gun barrel and fuselage. A flexible seal here was necessary since the gun oscillated 30-35 mm fore and aft with each shot. Note also the darker fuselage patch ahead of the Type 5; it is possible that this patch covers the ports for paired Type 99- 2s previously installed and removed. From a photo by ***James P. Gallagher***

A P1Y2-S *Kyokkou* night fighter at Atsugi in late 1945; this developed from the Frances. The painted-over tail code beginning in 3 speaks of previous service as a torpedo bomber. This Kyokkou differs from a typical specimen by having a single 30 mm Type 5 cannon substituted for the usual twin 20 mm Type 99-2 guns. The Type 5 may be installed slightly aft of the position used for the Type 99-2's and is slightly to port of the aircraft midline. ***Courtesy of James P. Gallagher***

Notes on Experimental Guns

A number of Pacific War era Japanese experimental aircraft machine guns and machine cannon are documented as having existed. For most of these, very little historical or technical data is presently available. The known experimental guns are therefore presented here with only brief notes about each, outlining the current knowledge of each.

Army Experimental Guns

Army experimentals are presented below in Te and Ho number order. It is very important to remember that Army Te and Ho numbers do not identify standardized gun types, but development projects. There were many Te or Ho numbered projects, some of which were never built, and only a few of which were formally adopted and given *shiki* numbers. Each Te or Ho number does identify one particular design, however. Prior to this designation system, gun projects were given simple descriptive titles, as shown in parentheses for the Ho-1 and Ho-3.

Te-000 series rifle-caliber guns: *(Two-digit designations are treated here as specific models of the single-digit guns, as they mostly appear to be.)*
Te-1: Flexible version of the Type 89 fixed, above.

Te-10 (poorly legible): Single, flexible, 7.7 x 58 SR.

Te-2: Three side-by-side barrels with one shared bolt, gas-operated, three box magazines, discontinued in 1941.

Te-26: Multi-barrel, flexible, 7.7 x 58 SR.

Te-3: Type 100/Type 1 flexible, above.

Te-35: (poorly legible) Same.

Te-4: See in operational guns, above.

Te-42: Single, fixed (sic), 7.7 x 58SR.

Te-5: Browning, fixed, 7.7 or 7.9 mm, dropped in initial study, 1942.

Te-6: Te-4 redesign dropped 1942. One source says "twin."

Ho-000 series, 20 mm
Ho-1 (20 mm flexible experimental): See in operational guns, above.

Ho-2: Fixed. May or may not be Type 97 based. Could be the Kayaba 20 (see below).

Ho-3 (20 mm fixed experimental): See in operational guns, above.

Ho-4: "Improved" version of Ho-1/3, believed to be belt fed, developed 1941-1942, dropped in favor of Ho-5. Kawamura design, probable predecessor of Navy Type 5 and 18-*shi*.

Ho-5: See in operational guns, above.

Ho-5-2: Flexible version of Ho-5, no details.

Ho-5-II: Lightened version of Ho-5, rejected as unreliable, 1943.

Ho-6 to Ho-11: References to 20 mm guns with these numbers are known, but no details. No testing reported. Most of these would be expected to use the 20 x 125 cartridge.

Ho-12: Browning gun for 20 x 125 cartridge, built and tested 1943-1945. There are photos, and an incomplete specimen was recovered and should exist in the U.S. This is the last design in this series, tested but unfinished at end of war.

Ho-050 series, 25 mm
Ho-51: Browning gun in 25 x 115, built and tested 1942-1945; photos exist, and a specimen recovered in late 1945 should be in the US. This design was the basis of the 30 mm Ho-155 prototype.

Ho-52: Browning gun in 25 x 150 (lengthened 25 x 115), tested.

Ho-100 series, 12.7 mm
Ho-101: Gas operated, fixed, 12.7 x 81 SR, tested and rejected 1941, possibly the Scotti ("Isotta-Mareschi").

Ho-102: Breda-SAFAT, 12.7 x 81 SR, tested and rejected 1941.

Ho-103: See in operational guns, above.

Ho-104: See under Ho-103 in operational guns, above.

Ho-150 series, 30 mm
Ho-151 to 154: Not mentioned in any documents found yet, these designations would be expected to have been used. Projects may not have resulted in actual guns.

Ho-155: See in operational guns, above.

Ho-200 series, 37 mm
Ho-201: Flexible automatic cannon. This might be the Kayaba design (see below)?

Ho-202: Fixed, may be a version of the Ho-201.

Ho-203: See in operational guns, above.

Ho-204: See in operational guns, above.

Ho-205: Browning, a reduced-size Ho-204, design started 1944, never completed.

Ho-250 series, 47 mm
Ho-251: Kokura Arsenal project developed 1943-5, never reached firing stage.

Ho-300 series, 40 mm caseless
Ho-301: See in operational guns, above.

Ho-400 series, 57 mm
Ho-401: See in operational guns, above.

Ho-402: 57 x 477 R tank/antitank gun with recoil operated automatic loader, built and tested 1943-1945. The Army-designed Ki-93-Ia twin-engined fighter was designed around this gun.

Ho-403: A little-known Kokura arsenal project, "in design stage" 1945, smaller and less powerful than the Ho-402, probable intended user of a reported 57 x 187 R cartridge.

Ho-500 series, 75 mm
Ho-501: 75 mm version of Ho-203, 1943-1945, in development at end of war—no testing reported.

Ho-502 to 504: No known documentary references.

Ho-505: 75 mm listed in United States Strategic Bombing Survey (USSBS) report, *Japanese Air Weapons and Tactics*, January 1947, as distinct from Ho-501. No other data. This may be the Type 88 antiaircraft gun fitted with a recoil-operated automatic loader.

Ho-600 series, 120 mm
Ho-600: Listed in above USSBS report, most likely synonymous with Ho-601 since a
Ho number ending in zero is unexpected.

Ho-601: "Machine cannon," likely an adaptation of an artillery piece to use an automatic loader. Rejected in design phase as impractical, 1944.

Ho-3000 series, caseless, various calibers
Ho-3057: 57 x 265 caseless cartridge Browning (not Oerlikon), single feasibility prototype with 5-rd box magazine now at USAF Museum, 1942-1945.

Ho-3075: 75 mm version of Ho-3057, abandoned in design phase, 1942.

Ho-3100 series, various calibers
Ho-3157: 57 x 187 RB, gas-unlocked blowback (Hotchkiss/Nambu based design), two unfinished feasibility prototypes; photos, both specimens recovered in late 1945, should be in the U.S. 1942-1945.

Ho-3175: 75 mm version of Ho-3157, abandoned in design phase, 1942.

Undesignated experimentals:
There were at least two "concept" experiments being conducted 1940-1945:

Engine-driven gun: A 12.7 mm gun operated by the aircraft engine, rather than forces internal to the gun, and very similar to an early Maxim-based Fokker patent was developed to firing phase, discontinued 1943.

Centrifugal gun: A non-firearm gun, which threw projectiles from a rapidly spinning wheel, was tested as a model only, 1941-1944.

MK 108: The Army contracted with Rheinmetall-Borsig in early 1945 for technical assistance with developing a license-built version of the MK 108, specifically listing the specialized springs required to run the gun. It was undoubtedly the Japanese inability to duplicate these springs under late-war conditions that kept this project from bearing fruit.

Other Army Guns
Chinn (*The Machine Gun*) reports a number of guns designed by one Shiro Kayaba, "the foremost Japanese designer of automatic guns." Kayaba is somehow rather less well known than Nambu (chief designer at Chuuou Kougyou) or Kawamura (chief designer at Japan Special Steel), despite his claimed prominence. Nonetheless, it is clear that his guns were built in prototype at least, and tested for the Army. There are two groups, a toggle locked recoil operated group, and a gas jet recoilless group using caseless cartridges or rockets. The calibers and other details suggest that all are later than 1939. They may all have had Ho numbers.

Toggle-locked
12.7 ("13") mm: Two versions are known from photos in Chinn: a belt fed (Vickers links) gun in which the toggle breaks upward, and a version using a 35-rd pan magazine in which the toggle breaks downward as in the larger-caliber versions. Both appear to use the 12.7 x 81 SR cartridge.

20 mm: Toggle breaks downward. The cartridge is not pictured in Chinn, but the Oerlikon-style drum magazine seems long for its diameter, so the cartridge is probably the 20 x 125, or perhaps even 20 x 142. Drum capacity is probably 30 rounds or less.

37 mm: Toggle breaks downward. The largest gun in this group uses a box magazine integral with the receiver, of about 5-rd capacity. The cartridge is pictured in Chinn, and is a powerful 37 mm round which does not match known Japanese types: it may be the 37 x 232 Italian 37/54 antiaircraft round. The gun in this photo may be a mockup and it seems unlikely that it was an aircraft gun, though it could be the basis of the Ho-201 or 202.

Recoilless group (these are very simple-looking guns):
40 mm: This gun was developed to the box-magazine fed firing stage. Cartridges were apparently similar to the Ho-301, but the magazine suggests a longer cartridge than the Ho-301. The gun reduced recoil by jetting propellant gas through a single nozzle at rear of gun.

80 mm: The photo in Chinn shows a hand-loaded single-shot prototype and cartridge. The cartridge seems to have a tail like the earliest Ho-301 rounds, and may be an out-and-out rocket, the gun a simple tube. This picture may actually show a test setup of the 74 mm ROTA antitank rocket gun, also referred to as the 7 cm *Funshin- Hou* (rocket gun), a bazooka-like infantry weapon, not an aircraft gun.

Navy Experimental Guns

Navy experimentals are listed here by *Kugisyo* development program.

The 14-*shi* program: The 14-shi program was initiated in 1939 when the 20 mm Oerlikon FF based guns entered service. It was a program in scaling the Oerlikon FF design down to produce a medium caliber gun.

14 mm 14-*shi*: Guns with this designation exist in the US in both fixed and flexible versions. The fixed gun may perhaps use a different cartridge, but the flexible gun uses a cartridge similar to the 20 x 100 RB Oerlikon L, necked down to 14 mm, with a slightly reduced body diameter to prevent accidental chambering of 20 x 72 RB rounds. The gun is very similar to the Type 99-1 flexible gun in all other respects. The fixed gun appears from its photograph in Chinn to be a Type 99-1 with a special barrel. Testing began March 1941 and was completed by July 1942.

The 17-*shi* program. In August 1942, the Kohon issued requirements for a "special large caliber machine gun" program. Several guns were developed under this program, most, if not all, disintegrating belt fed.

25 mm "Type 4": The Type 96 Hotchkiss antiaircraft gun modified for aircraft use, including a distinctive and effective muzzle brake, almost certainly belt fed. Heavy and slow firing, it was abandoned in favor of the 30 mm gun. A Yokosuka Arsenal project, this was apparently intended to meet the 25 mm 17-shi requirement below. The origin of the "Type 4" designation, from a U.S. intelligence document, is uncertain; "Model 4" (4 gata) seems more probable.

25 mm 17-***shi***: 25 mm gun using a less-powerful cartridge than the Type 96, no gun was designed exactly to this requirement.

30 mm 17-***shi-kou***: Two 30 mm guns were specified using the same *Kugisyo* designed 30 x 122 cartridge. This was to be 10 kg heavier than the *otsu*. No gun was developed to this requirement.

30 mm 17-***shi-otsu***: Type 5 30 mm fixed machine gun, in operational guns above.

40 mm 17-***shi***: This gun used a 40 x 150 cartridge known from specimens, and there must be a gun specimen in the US. Chinn diagrams it in great detail, but there is no photograph. It is very original, short-recoil operated with advanced primer ignition, rising-block locked, and disintegrating belt fed. Designed and developed primarily by Kure Naval Arsenal, it was rejected in favor of the 30 mm Type 5 gun in March 1945 after considerable development. Six projectile types, modifications of 40 mm Vickers types, were cataloged.

The 18-*shi* program. When the Kawamura belt feed for the Type 99-2 Model 4 was adopted in September 1943, a program was initiated to develop a faster-firing belt fed 20 mm gun. The result was the Type 99-2 Model 5, adopted in May 1945. One other gun from this program is cataloged in Japanese documents. A specimen exists in the USAF Museum.

20 mm 18-***shi***: Closely related to the Type 5, using a 20 x 98 cartridge. This gun may be a direct derivative of the Army Ho-4, rather than a scaled-down Type 5. The cartridge is a shortened 20 x 125, one version loaded with Type 99 projectiles but another having its own family of shorter, lighter projectiles. A *Kugishisyo* product, developed to successful in-flight testing, but discontinued March 1945.

Other Navy Guns
Italian 12.7 mm guns: The Breda-SAFAT and Scotti ("Isotta-Mareschi") medium machine guns were evaluated in considerable depth beginning in August 1939 and continuing through at least 1942.

The Breda-SAFAT underwent flight tests, and some work was done toward making the Scotti useful in flexible mode, including development of a saddle-drum magazine.

Oerlikon FFS: The third and largest Oerlikon FF-based gun was not forgotten. There was an ongoing program from March 1939 until at least September 1942 to improve the performance of the Oerlikon FFS as a *moteur canon* ("*moutaa kanon*" as rendered in Japanese) for "future use." Designations found that refer to the FFS are "*E shiki* 3 *gata*, *E shiki* 3 *gata kai* 1" (old-style nomenclature) and "*E shiki* 3 *gou* 2 *gata*" (new-style nomenclature), indicating Japanese production.

Service Branch Unknown

23 mm ?: A single short note from a British intelligence source has been found which indicates the evaluation of a 23 mm engine gun with a 120-round drum magazine on a Nakajima-built Hispano-Suiza V-12 aero engine in January 1941. This gun may have been an experimental Hispano-Suiza weapon, but it is not identified.

Glossary

Advanced primer ignition: in a blowback gun, firing of each shot while the bolt is still moving forward in closing, allowing use of a much lighter bolt and action spring, since the bolt's forward momentum compensates for lowered mass, and also somewhat counteracting subsequent recoil: in a short-recoil gun, firing of each shot while the bolt and barrel, already locked together, are still moving forward in counter-recoil within the receiver to counteract subsequent recoil.

B: in formal cartridge nomenclature, "belted," having a belted head like some rifle magnum cartridges.

Blowback: a gun operated by residual propellant gas pressure in the firing chamber: not locked, the bolt is kept closed during firing of each shot only by the mass of the bolt and the strength of the action spring.

Box magazine: a magazine in the form of a rectangular box, the cartridges stacked within, usually in one to two columns, over a follower pushed outward by a spring.

Buffer: a device used to dissipate or store excess energy.

Charger: a device for drawing back the bolt, or otherwise cocking or cycling the action preparatory to firing.

Disintegrating belt: a machine gun feed belt composed of usually metallic links that fall apart upon withdrawal of the cartridges.

Double drum magazine: two drum magazines joined side-by-side on top of the receiver with a common feedway between them.

Drum magazine: magazine shaped like a drum, its axis lying parallel to that of the gun, and the cartridges within it lying parallel to its axis, nose forward.

Falling-block breech: barrel breech closed by a vertically moving block that falls to open and eject each empty cartridge, rather than by a horizontally moving bolt.

Falling block locked: barrel and bolt locked together for firing of each shot by a block that moves downward with respect to the gun axis to perform this function.

Fixed: a gun not moveable as, say, to aim during normal use.

Flexible: a gun moveable for purposes of changing aim during firing.

Free: a flexible gun moved about entirely by hand without mechanical assistance or control.

Gas (operated): a gun operated by propellant gas drawn off from the barrel and acting on a piston.

Gas-unlocked blowback: a gun operated primarily by blowback, but held locked briefly to allow gas pressures to fall to safe levels and unlocked by a gas system.

***Giken*:** Army Air Technical Laboratory at Tachikawa, from May 1942 onward subdivided into the Air Testing Department and eight Air Technical Laboratories, with the Air Testing Department and Third Air Technical Laboratory sharing gun development work and combined in April 1944.

Kawamura belt feed: a belt feed designed by Dr. Masaya Kawamura of Japan Special Steel for the Type 99-1 and Type 99-2 Model 4 guns that attached to the magazine mounts and could be replaced by a magazine.

Kawamura double drum: double drum magazine, joined by a gear train ensuring that the drums unload alternately and in sequence, invented by Dr. Masaya Kawamura of Japan Special Steel.

Kjellman locked: the bolt is locked closed during firing of each shot by two lugs in the bolt sides that are rocked into position by passage of the firing pin striker between them. Invented in Sweden in the nineteenth century.

Kohon: *Kaigun Koku Honbu*, Naval Air Headquarters.

Kugisyo: *Kaigun Kokugijyutusyo*, the Naval Air Technical Arsenal at Yokosuka.

Kugishisyo: *Kugisyo Shisyo*, *Kugisyo* branch arsenal established at Kanazawa, after April 1941 responsible for design and development of all aircraft accessories, as opposed to the aircraft themselves, including armament.

Lewis drum: a drum-shaped magazine for the Lewis gun that may better be thought of as a multi-layered pan magazine, since its axis is vertical and the cartridges are arranged radially within (nose in).

Long recoil: a gun operated by barrel recoil motion equal to or more than the length of the cartridge.

Muzzle booster: a device affixed at the muzzle of a recoil-operated gun, utilizing Muzzle blast to intensify recoil effect on the barrel.

Muzzle brake: a device affixed at the muzzle of any gun, utilizing muzzle blast to counteract recoil.

Pan magazine: a flat, discoid magazine in which the cartridges lie radially, nose in.

Quadrant magazine: a magazine shaped like a broad wedge of a cheese wheel.

R: in formal cartridge nomenclature, "rimmed," a rimmed cartridge.

RB: in formal cartridge nomenclature, "rebated rim," a cartridge having a rim of less than case head (base) diameter.

Receiver: that part of a gun that "receives" (contains) the moving action and feed parts and the barrel of the gun, non-moving with respect to these other parts.

Rotating bolt head locked: the head of the bolt rotates to lock the bolt to barrel during firing of each shot.

Rotating collar locked: a collar surrounding both the bolt and the breech end of the barrel rotates to lock both together for firing of each shot.

Rising block locked: barrel and bolt are locked together during firing of each shot by a block that rises with respect to the gun axis to perform this function.

rpm: rounds per minute, shots fired in a hypothetical minute of continuous fire.

Saddle drum: a magazine composed of two equal drums hung at the sides of the gun and connected across the top of the receiver where the gun is fed.

Self-propelled: of a magazine, powered by the action of the gun rather than by an internal spring.

Short recoil: a gun operated by barrel recoil motion of less than the full length of the cartridge.

SR: in formal cartridge nomenclature, "semi-rimmed," having an extraction groove like a rimless case, but with a rim slightly larger in diameter than the case head (base).

Synchronized: forced to fire through the blades of a rotating propeller, usually not at the "natural" cadence of the gun.

Tilting bolt locked: the bolt is locked closed during the firing of each shot by the tilting of the bolt upward or downward in the receiver so that the rear end of the bolt bears against a locking face therein.

Toggle locked: the bolt is held closed during the firing of each shot by a straightened elbow-like linkage that must be bent to allow the bolt to open.

6

Gunsights and Bombsights

with assistance from
Ross Whistler & Shorzoe Abe

Optical systems for Japanese aircraft are a specialized field to themselves, and therefore warrant more than a passing description. It is in the optical field that Japan made the most notable advances during the Pacific War. This ranged from the fixed ring and bead, to gyroscopic lead computing gunsights and telescopic bombsights, yet none reached the level of technology that the Western powers had achieved.

Although relatively little has been published on this subject, it represents a major area of the technology found in Japanese aircraft equipment that deserves further description. The most detailed study in print is titled *History of Japan's Optical Industry with Special Reference to Military Equipment.* This was prepared in 1955 by the Compilation Committee, Japan Optical and Precision Instruments Manufacturers' Association. This interesting narrative was translated from the Japanese by Dr. Hajime Sakai and subsequently edited by Ross Whistler. This translation was the source of more than half of the historical and technical material contained herein relative to gunsights and bombsights. *(Note: The spelling of gun sight/gunsight seems optional. The Webster Dictionary does not list gunsight, yet most wartime manuals on this subject use gunsight. For consistency, the military preference will be used here.)*

The first optical gunsight used in the Japanese Navy was this Oigee/Aldis telescopic variety, replacing the ring and bead fixed sights. Japanese produced models were the Type 95 and Type 99. Shown here is the Type 99 Gunsight mounted on an Aichi D3A2 Val. ***80-G-345604***

Gunsights

During the Sino-Japanese war and the opening phases of the Pacific War, Japan had placed considerable emphasis on air-to-air dog fighting, not only for their single and twin-engine fighters, but for two-place carrier dive bombers and reconnaissance float planes, as well. Since these aircraft had the capability of dive bombing, the gunsight installed on them also served as a dive-bombing sight. Hence, the terminology of gun/bombsight was adopted for those instruments manufactured after 1935.

This is the viewing end of the Navy Type 95 Gun/Bombsight. To protect the forward lense, the lever is rotated 90 degrees to reposition the lense cover. Some installations have extensions to this leaver to pass through the windshield. ***Courtesy of Richard Lane***

Navy Gunsights

Oigee/Aldis Type Gunsights (Type 95 and 99)

When the Pacific War began, several aircraft were in service with obsolete-appearing telescopic gunsights, including the Aichi D3A Val and the Mitsubishi F1M2 Pete Observation seaplane, both having air-to-air combat capability. These early gunsights were imported from their original manufacturers, Oigee/Aldis. *Optische Anstalt Oigee* was a German firm, and Aldis Brothers was British. Eventually the *Nippon Kogaku Kogyo K.K.* (Japan Optical Co.) manufactured sights copied from these imports. The first of these was known as the Type 95 Gunsight, installed initially on the pre-Pacific War Aichi D2A1, Type 96 Carrier Bomber, Allied code named Susie, and later on the Aichi D3A1, Type 99 Carrier Bomber Model 11, Val 11.

The Type 99 Gunsight followed, installed on the later model Val 22. The Type 99 differed from the Type 95 Gun/Bombsight in having hermetically sealed optics, thus preventing internal fogging as altitude and temperature changed while in a diving attack. Although an improvement, it still proved unsatisfactory in the high heat and humidity of the South Pacific environment. In addition to having luminous paint on the external ring and bead sights, this instrument also had an illuminated optical aiming pattern for night bombing attacks. These sights had an effective aperture of 38 mm, focal length of 100 mm, magnification 1x, and 20 degrees field of view.

The Oigee type telescopic sight had several advantages over the earlier metal ring and bead sights, including greater eye freedom, primarily along the sight axis, but also laterally. Both the reticle and the target appeared to be in focus and superimposed at infinity. There was only one aiming element (the reticle image) to contend with, in contrast to having to align both ring and bead sights.

There were disadvantages, however. In addition to the tendency to fog and the exposure of the front lens to oil spray from the engine, there was inherent "tunnel vision," some light loss, wind

Japanese beautifully packaged all delicate instruments and equipment. The Navy Type 95 Gun/Bombsight pictured here is in its original box, with accessories that included a yellow screw-in eyepiece filter, leather eyepiece cover, screwdriver, dusting brush, two keys, and a silk cleaning cloth. ***Courtesy of Richard Lane***

Navy Type 98 Gun/Bombsight used on various Navy fighters, including a known sample found on a E16A Paul reconnaissance seaplane.

Type 98, left rear above. This is a damaged sight with glass reflectors missing and cut connector cable. ***Courtesy of USAAF via Ross Whistler***

resistance, and a hazard to the pilot's head in the event of a crash. When installed in front of the aircraft windshield the eyepiece was too far from the pilot's eye, resulting in an unacceptable reduction in the field of view. The optimum distance between the gunsight and the pilot's eye was only 102 mm.

The Japanese Naval Air Force (JNAF) believed that extending the distance of the eye to the lens to about 200 mm might work better, and requested the *Enomoto Kogaku K.K.* (Enomoto Optical Co.) in August 1935 to make an experimental model of such a modified gunsight. The end result, however, was disappointing and not approved for production.

Navy Type 98 Reflector Gunsight

This was the first of the reflector type (*kozo*, in Japanese) gunsights to be installed on Navy aircraft following the Oigee/Aldis fixed telescopic sights. The acceptance of this concept was very slow in coming, following the first testing of a French import in 1932, manufactured by O.P.L. (*Optique et Précision de Levallois*). One major and continuing factor was the safety hazard during an emergency landing, since the gunsight was even more massive than the telescopic model and was located directly in front of the pilot's face. This and other elements so concerned the JNAF officials that no further consideration was given to the reflector type sight for several years.

In the interim, the JNAF purchased anywhere from 12 to 30 (possibly 40) Heinkel He 112 fighters (depending upon the source of information stating quantity delivered) and spare parts in 1938, thinking they would have greater speed than the Mitsubishi Carrier Based Type 96 Fighter, Claude. This proved to not be the case, and the fighters were placed in storage, along with their installed Revi 2b (Revi is an abbreviation for the German term *Reflexvisier* (reflector sight in English)), their technology seemingly ignored. However, their design would reappear in the form of the Japanese reflector gunsight Type 98.

The JNAF realized that utmost urgency was needed to improve their fighter planes after the start of the Sino-Japanese conflict, and concluded that the telescopic gunsight could not possibly meet the requirements of future fighters. The suitability of the reflector type sight as the most likely replacement was soon realized. The Optical Department of *Kaigun Koku-Gijutsu-Sho* (Naval Technical Air Arsenal), better known as *Kugisho*, conducted a thorough test of the Revi 2b that came with the He 112 fighters, and saw a possibility of manufacturing a copied version by a domestic optical company. Of the anticipated technical problems, the greatest appeared to be the fabrication of the filter and the design of the f/1.2 optical system.

Solving these problems involved several groups: those engaged in gunsight development, in aircraft structural design, and in aircraft armament. Eventually they produced a successful gunsight that was first used in the well known Zero fighter, which was used in combat for the first time over China in mid-1940. The first test model of this Type 98 Gunsight became available about three months after mass production of the Zero was fully under way.

It soon became obvious to the JNAF that production of the new gunsight could not meet that of the aircraft for which it was intended. A decision was made to use the Revi 2b sights that were in the German He 112 fighters still in storage at Kasumigaura Na-

Left: Type 98, right side low. The protective crash pad at left was of rubber, leather covered, or rubber coated, depending upon manufacturer. ***Courtesy of USAAF via Ross Whistler***
Below: This is the Navy Type 98 Gun/Bombsight Model 1 circular reticle pattern for fighter aircraft. When used for dive bombing as the Type 98 Model 2, this pattern would have a grid image projection. This model was manufactured by *Tomioka Kogaku Keiki Seisakusho K.K. Courtesy of Todd Pederson*

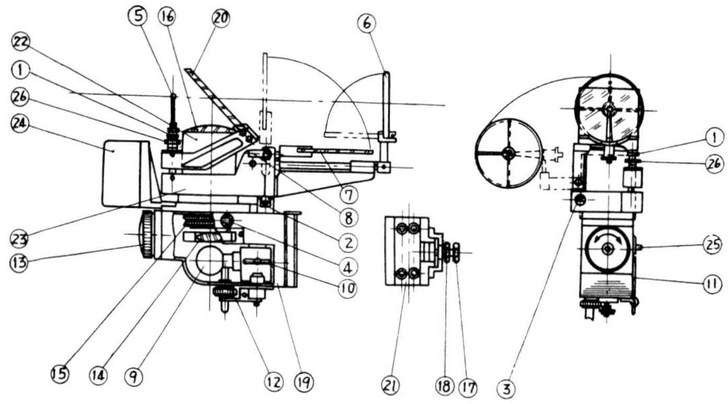

Navy Type 98 Gun/Bombsight Model 1

1. Elevation adjustment
2. Azimuth adjustment lock nut
3. Reticle lock screw
4. Reticle adjust screw
5. Auxiliary bead sight
6. Auxiliary ring sight
7. Filter.
8. Filter operating lever
9. Lamp
10. Lamp holder handle
11. Lamp access door
12. Lamp holder release
13. Rheostat
14. Condenser lens
15. Reticle
16. Collimating lens
17. Mounting clamp screw
18. Mounting lock nut
19. Lamp housing
20. Reflector plate
21. Mounting adapter
22. Collimator housing
23. Frame
24. Rubber crash pad
25. Condenser selector lever
26. Elevation adjustment and lock nut

val Air Base. The modifications made to these units consisted of a changed mounting plate and reticle. Thirty-two of the prototype Zero fighters were equipped with these Revi 2b Gunsights, while all subsequent aircraft were equipped with the Type 98 produced by the *Tomioka Kogaku K.K.* (Tomioka Optical Company). Aircraft types that are known to have been equipped with the Type 98 Gunsights were Zeke 21, 22, 32, 52, Rufe, Jack, Rex, George 11, Paul, Judy 11 and 43, and Irving.

The principal elements of the reflector gunsight are an illuminating lamp and housing, a reticle assembly, a collimator, image reflector, and filter (see illustration). The light rays emerging from any given point on the reticle are made parallel by the collimating lens, then reflected towards the gunner's eye by an optically flat diagonal plate of glass. To the gunner, the collimated reticle image appears to be at an infinite distance in front of his aircraft (but its angular size remains constant; it does not vary with distance). When the target lies along the line of sight the reticle image is superimposed upon the target, and both are seen as a single "sight picture." No parallax exists between these two images due to the action of the collimating lens, and only a single image of the reticle appears because the reflector plate is made of plane parallel glass. If the reticle image is not properly collimated, or if the two surfaces of the reflector plate are not parallel, a double image of the reticle pattern will appear. Beyond the requirement for proper collimation and a plane parallel reflector plate, an aircraft gunsight should be able to produce a bright enough reticle image to show up against a cloud background, and should have a large field of view in order to display a large enough reticle to accommodate the substantial lead angles encountered in high-angle off shooting. Target range is usually short—about 500 meters or less. The reflector gunsight satisfies these requirements much better than the telescopic gunsight.

Because of the protruding nose of single-engine bombers, over-the-nose vision was limited to 5 degrees when looking through the original 1937 reflector type gunsight. The earlier telescopic sight had afforded a much larger downward sighting angle of 12 to 13 degrees, since its entrance pupil, or objective lens, was some 800 mm closer to the nose of the aircraft. In dive bombing and high deflection shooting a downward field of view over the nose of more than 10 degrees was required to permit sufficient line of sight depression to keep the target in sight. To achieve this with a reflector sight, it would be necessary to raise the sight as well as the pilot's seat. This would result in more frontal area and increase total drag, not fully offset by the elimination of the externally mounted telescopic sight. A compromise resulted in a downward field of 7.5 degrees through minor design changes, mostly in sight positioning. (Pilot note: It was not uncommon in some post war aircraft using reflector type sights for glide bombing—such as the USAF Martin B-57 and North American F-100—for the target to be tracked passing under the nose and out of sight to the pre-determined bomb release point of the depressed pipper, as experienced by the author.)

Two models of the Type 98 Gun/Bombsights were manufactured. Model 1 was mainly installed as a gunsight on fighter-type aircraft. It had a circular reticle pattern. Model 2 had a square grid reticle pattern and was installed on carrier aircraft used primarily

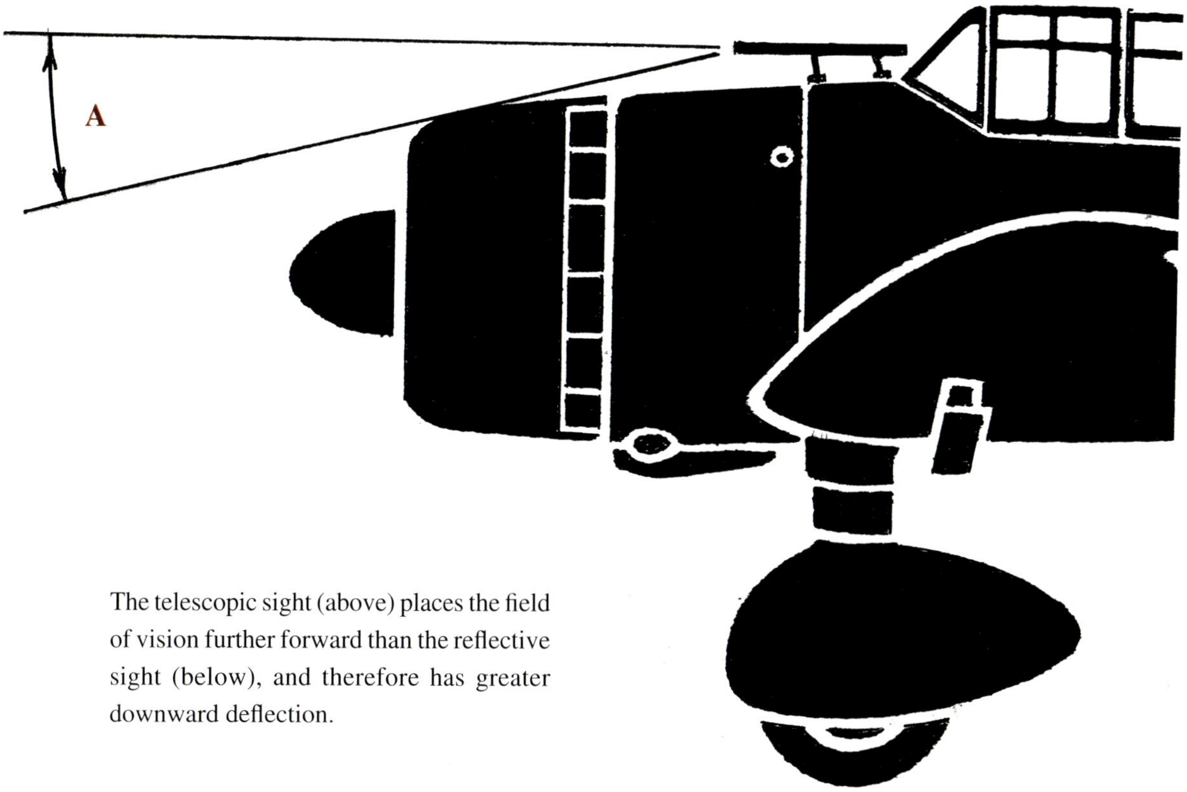

The telescopic sight (above) places the field of vision further forward than the reflective sight (below), and therefore has greater downward deflection.

Gun/Bombsight angular deflection restrictions for telescopic and reflector sights.

Navy Type 2 Mk.1 Gun/Bombsight Model 1. This sight was used well into the war on dive bombers such as Judy as a replacement for the Type 99 Telescopic Sight. A rotating prism in the enlarged nose adjusted deflection angle for use as a gunsight or, when changed by computer input, provided a desired deflection angle. ***Courtesy of Todd Pederson***

for dive bombing. Model 1 sights were manufactured at *Tomioka Kogaku K.K.* at a rate of 1000 per month, and at *Chiyoda Seimitsu Kogaku K.K.* (Chiyoda Precision Optical Co.) at a rate of 200-300 per month. Model 2 sights were assembled mainly at *Nihon Kogaku Kogyo K.K.* and *Tokyo Kogaku Kikai K.K.* (Tokyo Optical Co.).

Navy Type 2 (Telescopic) Gun/Bombsight

There were those in the Navy that felt there was a continuing need for the telescope Gun/Bombsight, despite the advantages of the reflector sights. The Navy Type 2 Mk.1 Gun/Bombsight Model 1 was a development of the telescopic concept. The designation Type number 2 refers to the Japanese year 2602, or 1942 in Western usage.

This sight had a long development history that can be traced to 1937, when the JNAF purchased the Heinkel He 118 V4 prototype in order to learn about Western state of the art for dive bombers for use in the development of the Kugisho D4Y1, later to be known as Judy. An item on the Heinkel that drew particular attention was the gun/bombsight. This instrument required altitude and airspeed inputs, and dive angle obtained from a gyroscopic instrument operated by the crewman in the rear seat. This instrument was called the Gyro Stabilized Inclinometer Model 1 (see illustrations of this unit installed in Judy.). Its purpose was to control the positioning of the reflector plate of the pilot's gun/bombsight by measuring the angle between the horizon and the aircraft's longitudinal axis. To this angle had to be added the aircraft's angle of attack in order to produce the true dive angle of the aircraft, and therefore the angle of release of the bomb. Various other factors had to be taken into account, notably the ballistics of the specific bomb being dropped, in order to derive the correct depression of the line of sight. Presumably all these variables were calculated during mission planning, and the necessary values of certain inputs were set into the computer, which then automatically calculated the correct bomb release point and sight line depression angle. From this computing unit, a series of cables and pulleys drove a mechanism mounted inside the aircraft under the forward mount for the sight. This system automatically adjusted the off-set dive angle and bomb release point for striking the target. Further development was undertaken by the Instrument Department of Kugisho, which made improvements upon this concept that increased bombing accuracy. So promising was this new developing technology that an all new design was begun.

The new design continued to evolve around the gyro stabilized inclinometer, one that synchronized and rotated a prism in the forward end of the telescope. This accounts for the enlarged nose on the Type 2 Gun/Bombsight.

The advantage of this sight over previous types was that limited training was needed for the deteriorating abilities of fledgling pilots. The pilot only needed to align the target in the center of the depressed sight reticle, adjust for windage, and the off-set dive angle was pre-computed and fed into the sight.

These telescopic sights were used in Navy dive bombers such as Val, Judy 12 and 33, and early installations for Seiran. This Type

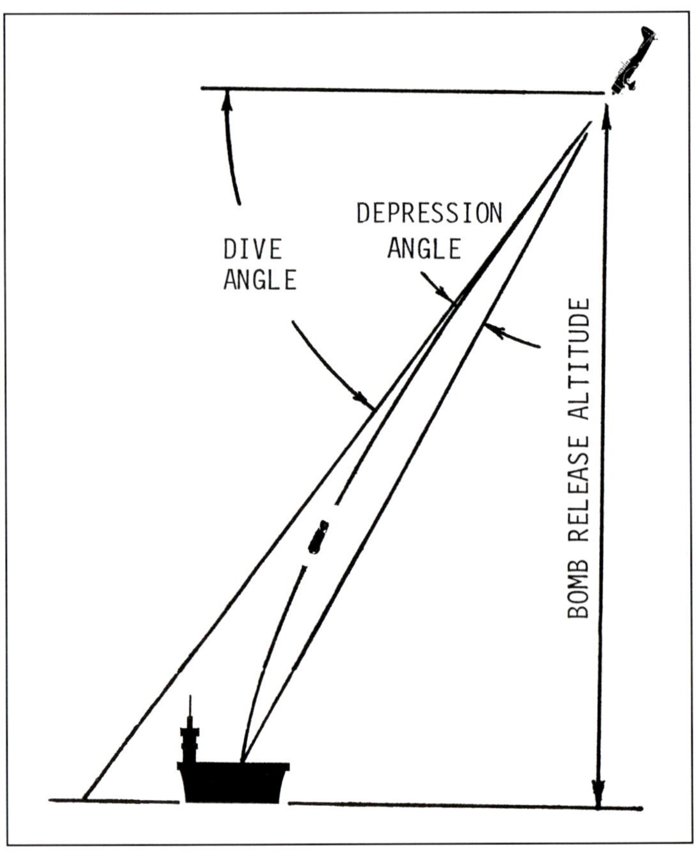

Gun/Bombsight depression angle for dive bombing.

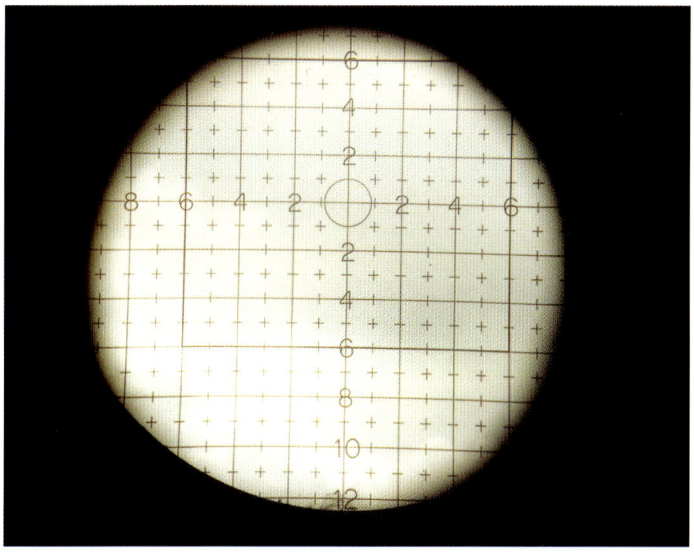

Navy Type 2 Gun/Bombsight reticle image used for Navy dive bombers. A grid pattern is used instead of a circular reticle pattern for fixed gunnery. ***Courtesy of Todd Pederson***

2 Gun/Bombsight served for dive bombing as well as gunnery. Production ended in July 1943.

Of interest with this sight is the bullet-shaped nose cover to the telescope that protected its optical surface. When the sight was to be used the pilot manually rotated a torque rod, running from the eyepiece end of the telescope to the cover, through 90 degrees. This moved the cover to the side while the sight was in use. The placement of these sights outside the aircraft reduced the aircraft speed by about 3-4 knots at 200 knots.

Navy Type 3 (Reflector) Gun/Bombsight

As an armament system, this sight operated very much like the Type 2 Mk.1 Gun/Bombsight. Its main difference was in using a reflector sight instead of a telescopic, as well as having an updated computing unit. In this case the sight angle was adjusted by a three dimensional cam that compensated for the difference between the dive angle and the line of trajectory. As with the Type 2, the Type 3 required a Type 1 Control Unit (for Gun/Bombsight) Gyro Stabilized

This is the forward end of the Navy Type 2 Gun/Bombsight, which contains the adjustable prism for sighting. This view shows the lens protector rotated to the side when the sight was in use. ***Courtesy of Todd Pederson***

Another view of the Navy Type 2 Gun/Bombsight nose shows the calibration drive wheel at the bottom mounting location that plugs into the drive mechanism within the aircraft. This particular sight in the NASM collection was manufactured by *Nippon Kogaku Kogyo K.K.* in January 1945. ***Courtesy of Todd Pederson***

Navy Type 3 Mk.1 Gun/Bombsight Model 1. This large and very complex sight appeared late in the war. ***80-G-191854***

The recently restored NASM *Seiran* submarine based twin floated bomber was originally equipped with a Navy Type 2 Model 1 Telescopic Sight, as evidenced by the plugged windshield. Provisions for mounting the Navy Type 3 Model 1 Gun/ Bombsight were found, so this example was taken from storage for installation in the airplane.

This is a Navy Type 3 Mk.1 Gun/Bombsight Model 1 with many vital parts souveniered. This appears to be a model variation of the standard unit. It is shown here mounted in the Aichi B7A1 Grace that is in the National Air and Space Museum awaiting restoration.

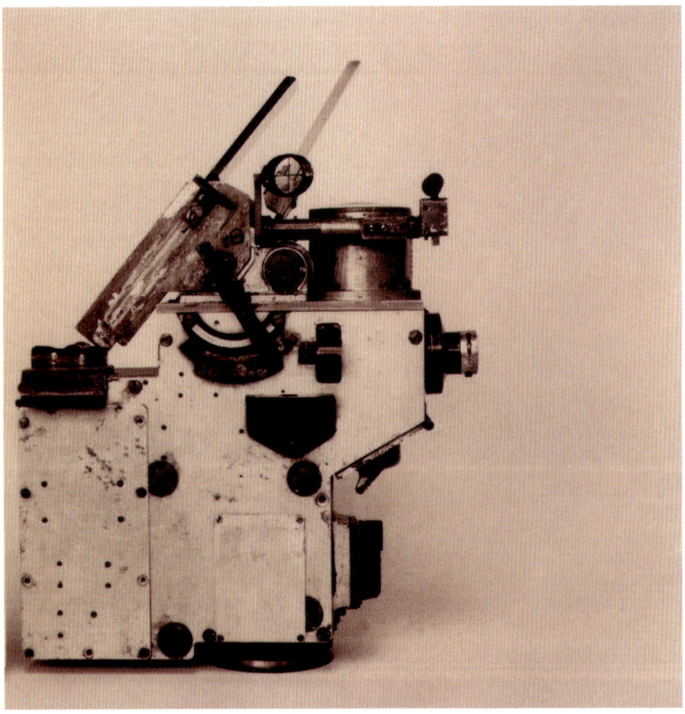

Above: Left side of Type 3 Mk.1 Gun/Bombsight Model 1. This type of sight was also used in Frances and *Seiran*. *80-G-191851*

Right: Pilot's view. *80-G-191852*

Top view. ***80-G-191853***

A vital part to the Type 3 Mk.1 Gun/Bombsight Model 1 is this Type 1 Control Unit (for Gun/Bombsight) Gyro Stabilized Inclinometer. Mounted at eye level in front of the rear seat, the observer inputs certain factors that through a cable linkage automatically sets the bombsight for the proper release point and dive angle. Manufactured by ***Tokyo Koku Keiki. 80-G-192299***

Navy Type 3 Compact Sight. This sight was especially designed for night fighters when using the obliquely mounted cannon for firing upward at a 30 degree angle (see Irving section that shows installation). By necessity these were small, since they had to be mounted above the pilot's head on the windshield. ***Courtesy of USAAF via Ross Whistler***

Inclinometer in the rear seat for the observer to manage adjust the Reflector Plate unit inside the sight.

Accepted by the Navy in January 1944, this Type 3 was first installed in the Kugisho P1Y1 *Ginga*, code named Frances. A Type 3 Mk.1 Gun/Bombsight was specified for Seiran conversions, and a Type 3 Mk.1 Gun/Bombsight Model 1 with slight differences was installed in Grace; examples of both are mounted respectively in National Air and Space Museum aircraft.

Navy Type 3 Compact Gunsight

The effectiveness of a weapon system is only as good as its weakest major component. In the case of the Japanese Navy's development of upward-firing night fighters for attacking enemy bombers, a special gunsight had to fit into a small area behind the upper portion of the fighter's windshield. To fill this requirement, the Type 3 Compact Gunsight Model 1 was developed.

The Japanese need for revised night fighter tactics became painfully apparent at Rabaul, New Britain, which came under attack almost nightly by American B-17s. When the commander of the operations at Rabaul (Commander Yasuna Kozono) returned to Japan, he took with him a concept that he wanted to explore. Having achieved earlier success by dropping phosphorous bombs from an airplane into a formation of B-17s, he considered the possibility of mounting downward-firing guns on an aircraft that would do the same thing. Then, too, if this would work in firing downward at a target, it could fire upward, as well. When firing upward, the attacking aircraft would be in the darkest part of the night sky, while the target aircraft would be higher and silhouetted against the starlit night sky. With this form of attack, there was no need for calculating a lead on the target when firing the guns.

Early in 1943, Kozono's idea was accepted for trial only until proven, using three reconnaissance versions of the Nakajima J1N1 *Gekko* known as Irving. While the configuration changes were being worked out for installing the obliquely mounted 20 mm machine guns, there arose the need for a gunsight for this unusual firing configuration. Commander Kozono took this problem to Mr. Furuzawa, division head of Kugisho's Optical Department in Kamariya, Yokohama. To cooperate in this endeavor, regardless of the negative feelings of JNAF headquarters, Furuzawa discovered an inventory of about 200 German L-shaped reflector gunsights, several of which were already in his department for tests. These units were designed to be used with a remote control turret gun.

This gunsight, the Revi 25B, was manufactured by Steinheil Co. Its collimator had an aperture of 45 mm, and its field of view was 15 degrees. The optical system was arranged in an L-shape, which helped to keep its longest dimension quite small and out of the way of the operator. A test conducted with a modified reticle showed excellent results, and the sight was officially adopted under the designation of Navy Type 3 Compact Gunsight. The JNAF officially ordered Kugisho to begin the modification of these existing sights. The supply of modified sights was quickly exhausted as production of the J1N1-S Irving night fighter increased, and as Kawanishi's conversion of the twin-engine bomber P1Y1 Frances to night fighters progressed. The Koishikawa plant of *Riken Kagaku Keiki Seizo K.K.* (Riken Scientific Instrument Manufacturing Co.) undertook mass production of this gunsight as demand for the instrument increased.

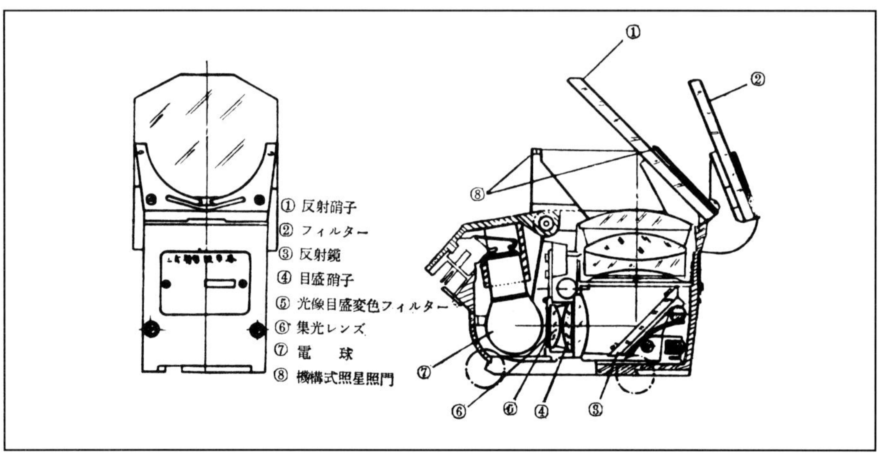

Navy Type 3 Compact Sight.
1. Reflector plate
2. Filter
3. Mirror
4. Reticle
5. Colored filter
6. Condenser lens
7. Electric lamp
8. Auxiliary ring and bead sight

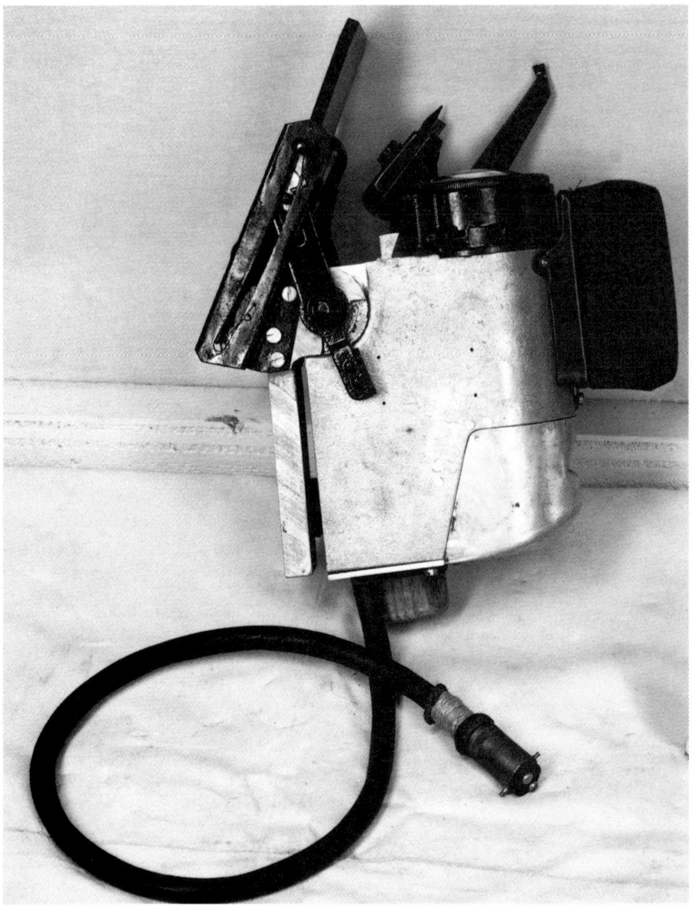

Navy Type 4 Gunsight Model 2. Based upon the Revi 12C design, this improved model developed in 1942 partially replaced Navy Type 98 Gunsights. The Army had a similar version emanating from the German design known as the Type 3. ***Courtesy of USAAF via Ross Whistler***

As it turned out, this method of night fighter tactics against Allied bombers became very effective. Aside from the concept itself, much of this success came from this adequate gunsight, although considered by some as awkward and cumbersome.

Navy Type 4 Gunsight (17-*Shi* Experimental Gunsight)

Toward the end of 1940, one of three Junkers Ju 88s ordered by the JNAF from Germany arrived in Japan. A reflector gunsight, the Revi 12C on the Ju 88, was an improved version of the Revi 3C; well suited for mass production because of fewer parts and a simplified design. This sight was widely used to great effect by Luftwaffe fighters during the first two years of the war, both on the single engine Bf 109E and the twin-engine Bf 110 and Me 210.

With this gunsight in mind, Kugisho engineers decided to improve upon the Navy Type 98 Gunsight then in widespread service. Lieutenant (Engineering) Masayoshi Shiraishi and Engineer Mikio Shirota, designer of the Type 98, collaborated in the design at the Aircraft Weaponry Department of Kugisho. Revised drawings were submitted to the *Tomioka Kogaku K.K.* and *Chiyoda Seimitsu Kogaku K.K.* for prototype models, allowing them to change details as they deemed appropriate. Their modified designs were approved for limited production in January 1943, but did not get underway until the summer of that year.

Comparing the two, the one produced by Tomioka Kogaku (designed by Engineer Takahashi) came closer to meeting the official military specifications. Based on this prototype and with modifications added, the Ofuna plant of the *Tomioka Kogaku K.K.* received an additional contract for a second prototype, which they completed in the spring of 1944.

Kugisho tested this product, and the results were very satisfactory; hence, the Navy Type 98 Gunsight was replaced by the improved version, Navy Type 4 Mk.1 Gun/Bombsight Model 1, which went into mass production in 1944.

However, when the B-29 raids started over the home islands of Japan, defending fighter pilots frequently were opening fire before they were within effective firing range. Since the B-29 was much larger than other 4-engined bombers, pilots tended to underestimate their size in relationship to their distance, and expended ammunition without effect.

Planners realized that it was necessary to make a gunsight capable of measuring target range based on a known target wingspan. It was recalled that the Brewster Buffalo fighters captured in Java were equipped with an optical gunsight having a variable reticle. This was the British Gunsight Mk II, which had an adjustable gap in the middle of the horizontal cross hair that could be set to frame a given wingspan at a desired firing range. Using this principle, a prototype Navy Type 4 Gunsight Model 3[1] was created at the Ofuna plant of *Tomioka Kogaku K.K.* in October 1944. This sight

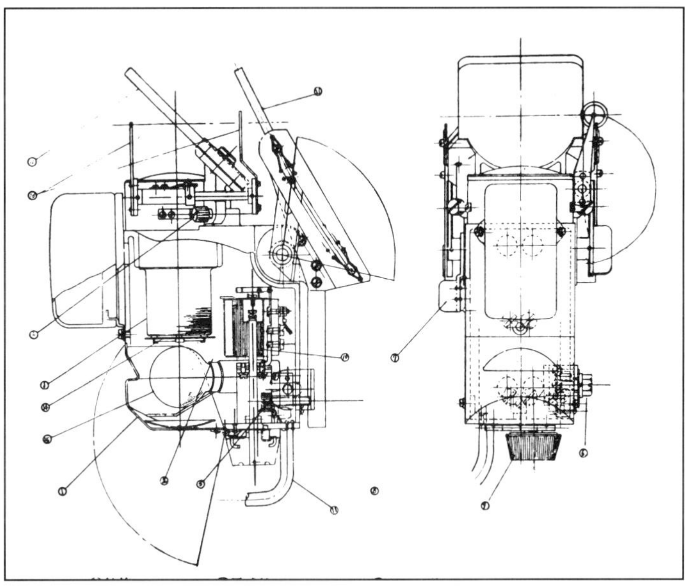

Navy Type 4 (17-*Shi*) Gunsight.

1. Lens barrel
2. Lamp socket
3. Mounting base
4. Azimuth adjustment
5. Vertical adjustment
6. Set screw for 5
7. Filter handle
8. Lamp cover
9. Rheostat knob
10. Rheostat
11. Electric cable
12. Reflector plate
13. Filter
14. Image filter
15. Electric lamp

Rear view from the right showing the earlier version Type 4 Gunsight Model 1. Note that the door to conceal the light missing from its socket is open. ***Courtesy of Todd Pederson***

Left side of Navy Type 4 Gunsight Model 1. ***Courtesy of Todd Pederson***

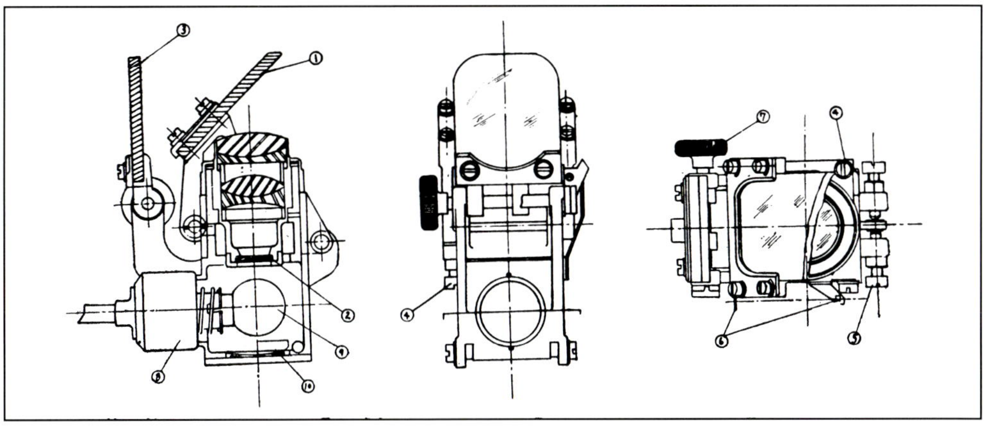

Navy Type 4 Compact Gunsight
1. Reflector plate
2. Reticle
3. Filter
4. Elevation adjustment
5. Azimuth adjustment
6. Auxiliary sight
7. Filter lever
8. Lamp socket
9. Lamp
10. Lamp condenser mirror

was evaluated and found favorable by the Flight Test Department of Kugisho, and was placed into mass production in 1945. This sight appeared on such fighters as Jack 21, 31, and 33, George 21, and Zeke 62 and 63.

Navy Type 4 Compact Gunsight

During the summer of 1940, Warrant Officer Mochizuki of the floatplane unit of the Flight Test Department requested the Aircraft Weaponry Department of Kugisho to make a prototype optical sight that would be suitable for flexible gunnery. He reasoned that the ring and bead sight was too primitive for modern air-to-air combat. An optical type was obviously better for this purpose, and he asked for a telescopic type gunsight to evaluate. The specifications given the Optical Department called for a small telescope gunsight with an aperture of 20 mm, a focal length of 200 mm, a field of view of 20 degrees, and having a magnification of one. The Department clung to their opinion that employment of such a sight was difficult because of its short eye to lens distance, and that internal fogging constituted a serious operational problem that they believed to be insoluble. Nonetheless, a prototype was made according to the specifications using a lens taken from the aiming sight of an Aldis signal lamp. The finished instrument was delivered to Kugisho for testing.

The test results confirmed the predictions of the Optical Department. The sight was useless because of fogging of the optics. The Aircraft Weaponry Department insisted that the instrument be completely immune to fogging, however, the Optical Department replied that this was impossible, as the telescope tube could not be made airtight because of its adjustment mechanism. The Optical Department's recommendation to use a reflector type sight instead was finally agreed upon, and the decision was reached to produce a prototype using available components. Two months later the prototype was completed, and firing tests began. Much to their disappointment, the sight failed to maintain alignment when the gun was fired because of the structural weakness of the mount. Efforts to strengthen the mount proved futile.

After several failures, it was decided in April 1942 to completely redesign the sight along the reflector sight principle, but to use a cast metal body. The Enomoto Kogaku K.K. was asked to make a prototype. That September the third modification finally satisfied the rigidity requirement. Results were so satisfying that at this point Kugisho wanted to completely convert to this type of reflector sight.

In reality, a complete conversion was almost impossible within a short period, since an enormous number of sights was required. Priority was given to reflector sights for the 20 mm machine gun mounts in the nose of the Kawanishi-built P1Y1-S Frances. Sight tests with this 20 mm machine gun were not as successful as those with the lighter 13 mm machine gun, revealing that further strength-

Navy Model 2 Reflector Flexible Gunsight Mk.1 *Ko*. These sights were mounted on flex-mounted machine guns and in turrets. This was a breakaway from the traditional ring and bead sight, although this sighting capability was retained as part of this instrument. *Nikko* was the manufacturer of this unit. *80-G-192675*

Unimpressive looking in this view, this Navy 18-Shi Experimental Gunsight was an adaptation of the Sperry computing gunsight acquired from downed B-17s. This unit gave lead sighting capability to bomber crews with flex-mounted guns. This is the front, facing the gunner.

ening of the mounting structure was needed. Official acceptance came in March 1944, after the necessary structural improvements were completed. This sight, designated the Type 4 Compact Gunsight, was mass produced by the *Enomoto Kogaku K.K.* (which was later acquired by the *Fuji Shashin Kogaku K.K.* (Fuji Photo Optical Co.)).

Navy Model 2 Reflector Flexible Gunsight Mk.1 *Ko*

The need became obvious that smaller and lighter weight flexible gun tracking sights were necessary. Improvements brought about several new models, one of which was the Model 2 illustrated here. These replaced the ring and bead sighting methods used for flex-mounted weapons.

Navy 18-*Shi* Experimental Gunsight

When the Japanese forces moved on to Malang Airfield in Java, Dutch East Indies, in April 1942, they captured two American Boeing B-17E Flying Fortresses. In this bomber was a Sperry computing gunsight used in the .50 cal turrets, one of which was sent to the Navy's Optical Department of Kugisho for evaluation. The sight contained a compact analog computer for lead calculation and adjustment, similar to the type used with antiaircraft guns. This mechanism impressed the departmental management, which eventually, in March 1943, asked the *Nippon Kogaku Kogyo K.K.* to make a reproduction.

The Japanese Army also became interested in this instrument and asked the same company to make a prototype for them, but one that differed slightly from the original, in that it would use metric rather than English dimensions. The Japanese Navy, on the other hand, anticipating possible complications when converting the analog computing mechanism to metric, decided to stay with the original dimensioning. Modifications were nonetheless required because of the difference in ballistics between the American .50 cal and the Japanese 20 mm ammunition.

The prototype was completed at the end of 1943 and, after ground testing, was declared successful. The tests showed that the computed lead angles were within 30 seconds of arc, or half a degree, of the theoretical value. However, when preparing the sight for flight testing, the power gun mount could not correctly mate with the prototype sight unit. Original plans called for this sight to be used with the upper and lower guns on the Nakajima G8N1

This is the right side of the Navy 18-Shi Experimental Gunsight. The protrusion at the top houses the reflective glass and lense for sighting.

Renzan long range heavy bombers—know as Rita—much like the turrets used on the American B-17. However, this installation became impractical for the lower turret, since the gunsight severely restricted the gunner's vision. Therefore, the installation was to be used in the upper turret only.

The G8N1 *Renzan* prototype was not completed by the time ground testing of the gunsight was finished and flight testing was to begin. Fuel shortages prevented the use of other aircraft for the in-flight testing phase. Despite this lack of complete testing of the prototype, the sight was put into production based upon the positive results of tests already completed. Production was turned over to the manufacturer of the prototype, *Nippon Kogaku Kogyo K.K.*, at their Kawasaki plant (near Tokyo).

Although committed to production, the third G8N1 *Renzan* four-engine bomber, finished in March 1945 and being flown at Misawa Naval Air Base in northern Honshu, was to flight test this gunsight in conjunction with other phases of aircraft testing. Because of engine problems and air raids, the 18-Shi Gunsight to be tested with the airplane was not completed before the war's end. The Army's Mk.2 was also unfinished, as both were of the same design principle and manufactured by the same company.

Navy 19-*Shi* Experimental Compact Gunsight

As experience was gained with the new concept of obliquely mounted guns for night fighters, the Type 3 Compact Gunsight was found to be too large for single-seat fighters. In August 1944 another type of reflector gunsight was developed that was even smaller. Its optical system employed an L shaped collimating system of 20 mm aperture. Since this gunsight was only to be used at night, the lamp was limited to a maximum brightness of only 5 candlepower. These gunsights were manufactured at the Ofuna plant of the *Tomioka Kogaku K.K.* The factory workers nick-named this sight the "*Mouse*," since the electrical lead connected to the tail end of the L-shaped body gave them the appearance of such a rodent. Mass production for this sight began in December 1944. Several non-critical components were made of wood in anticipation of shortages of

Officially this was the 19 Shi Experimental Compact Gunsight, better known by those associated with it as the *Mouse* due to its appearance. Production was intended for night fighter conversions for oblique firing fighters that had limited cockpit space. ***Courtesy of USAAF via Ross Whistler***

Production for the *Mouse* began in December 1944. Due to anticipated shortages of metal parts, several non-critical components were made of wood. Since the heat generated by this sight was quite low, its accuracy did not suffer from this substitution of some wooden parts. *Courtesy of Todd Pederson*

metal parts. Since the heat generated by this sight was quite low its accuracy did not suffer at all when some wooden parts were used.

This sight was planned for installation on the night fighter versions of the Mitsubishi J2M3 *Raiden*, Allied code named Jack. However, the war ended before this sight was used in combat.

Navy Prototype Gunsight Employing Angular Rate Measurement

When making a high-angle gunnery pass, the fighter pilot must make a continuous mental estimate of the correct lead angle so that the bullets will be at the same point and time when reached by the target aircraft. This mental calculation requires long and arduous training. The JNAF learned from an intelligence report obtained from Germany in 1943 that the Allied Air Forces had a gunsight that solved this triangulation problem by use of a rate gyro. The report speculated that the Allies' successful defense against German air raids over Great Britain was owed in part to the effectiveness of this sight.

Early in 1944, a Mr. Tada of the *Yokokawa Denki Seisakusho* (Yokokawa Electric Instrument Co.). proposed a workable solution to this tracking problem. He obtained his idea from a Sperry computing sight for turret guns on the previously mentioned captured B-17E Flying Fortress. According to Tada's thought process, the motion of an entire aircraft tracking a target is no different from that of a turret gun doing the same. He applied the operating principles of the Sperry sight to a device for guiding a fighter into firing position. Assume that the fighter is brought to bear directly on the target. At this point, if the angular turning rate is measured using a rate gyro, the desired lead angle is obtained by multiplying this angular rate by the time interval required for the bullet to travel to the target. A correction must also be included into this equation to account for gravity drop in the bullet's trajectory. If the gunsight generates an automatic correction for this angle, the pilot can hit the target by centering the target in an offset reticle pattern. Since the aircraft motion is two-dimensional in this problem, two angular rates—one in elevation and one in azimuth in the fighter coordinate system—must be known for determining the correction angle.

Tada's idea was submitted to both branches of the military; the JAAF's 5th Laboratory, and to the Aircraft Weaponry Department of the JNAF's Kugisho. The response from both services was swift, requesting fabrication of two working models within several months.

The linkage between the sight head and the rate gyro was by means of a pneumatic coupling. The Optical Department had extensive exposure to this type of instrument through their experience with the Type 2 and Type 3 Gun/Bombsights. Therefore, they had detailed knowledge of the instrumentation required for this gunsight and strong confidence in their capabilities. After having reviewed the test results on the second and third prototypes the JNAF headquarters ordered 30 units to be produced, and the Optical Instrument Department of the Ministry of Munitions[2] oversaw the production of these instruments. To be sure of satisfying all requirements of the military users, the Department organized, in early January 1945, a joint group consisting of the responsible personnel in the JAAF, the JNAF, and Mr. Tada of the *Yokokawa Denki Seisakusho*. The result was a request for delivery of two copies, by the end of February, of the third model, which would include several modifications, and of three more gunsights soon after. The decision regarding production of the remaining 25 units would depend on the results of the planned testing. The first unit was completed in the middle of March, and the second at the end of the same month. One unit went to the JAAF, the other to the JNAF, and the Optical Department took primary responsibility for testing.

The delivered units demonstrated a substantial lag in the servomechanism response, as well as other minor problems. Flight test was initiated after extensive modification and adjustments. The first unit was installed on a fighter at Atsugi Naval Air Base, and the second on a Kugisho Flight Test Department aircraft at Misawa Naval Air Base, located there under the JNAF dispersal program. The flight tests were disappointing, as the compact and delicate mechanism repeatedly failed. This gunsight project was not completed before the war ended, and the sight was never used in combat. However, the rapid progress demonstrated in the development of this epoch-making gunsight was due to collective cooperation between the two services, which were often in conflict with each other.

Army Gunsights

In the early days of Japanese Army Air Force history, the gunsight used was the bead and ring type. As aircraft were improved, this sight was replaced by a simple telescopic gunsight, which was a low power monocular having a cross hair in its reticle plane. These sights were either of the Chretien (French) or Oigee (German) type. The official name was Type 89 Sighting Telescope for Fixed Machine Gun.

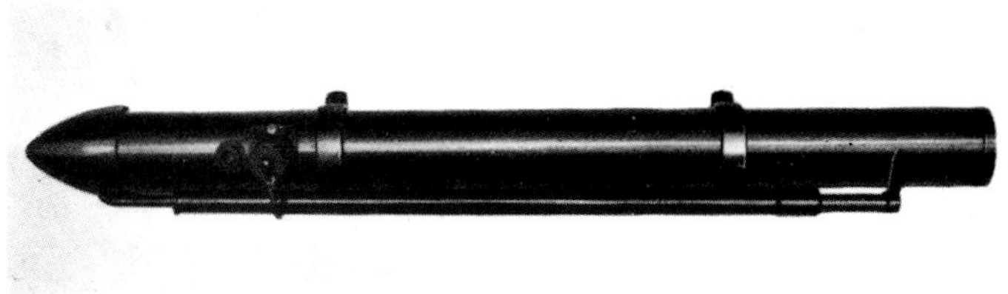

Army Type 89 Sighting Telescope.

Army Type 89 Sighting Telescope

This Army Type 89 Sighting Telescope, or telescopic gunsight, gave more eye freedom, as described for the Navy's version of this Oigee type gunsight. Maintaining a fairly exacting aiming position for the pilot's eye in order to bring the target into the field of view remained difficult, however, especially while maneuvering the aircraft. Despite the addition of a soft rubber eyeguard to help steady the pilot's eye, the task became harder as aircraft performance improved appreciably in the late 1930s. Oddly, when compared to Japanese Navy standards at that time, early models of the Ki-43 Oscar, Ki-44 Tojo, and Ki-51 Sonia were equipped with this telescopic gunsight, while the Navy's Zero, which preceded these fighters, was equipped from the beginning with the Navy Type 98 reflector gunsight.

The Japanese Army entered the Pacific War while still using the monolocular telescopic gunsight. Shown here is a Nakajima Ki-44 Tojo with its Army Type 89 Sighting Telescope protruding through the windshield to bring the sight closer to the pilot for easier sight alignment.

A close-up cockpit view of a Nakajima Ki-43-I shows the rubber eyepiece of the Army Type 89 Sighting Telescope. Note the toggle lever for moving the front lens shield to the side.

Army Type 100 Gunsight

The JAAF Air Research Institute at Tachikawa first considered a reflector gunsight in 1939 when they asked the *Tomioka Kogaku K.K.* to develop a prototype. It differed from its Navy counterpart Type 98 in utilizing an "L" shaped optical design. Viewed as being rugged in basic design, it was also quite large for a fighter sight. Test results were gratifying, however, and production was expanded to *Nikko* and *Tokyo Kogaku Kikai K.K.*, entering Army service in 1942 as the Type 100 Gunsight, and replacing the Type 89 telescopic sight. This became the most widely used Army reflector gunsight of the war, being used in such aircraft as the Oscar 2 and 3, Tony 1, Tojo 2, Nick, Randy, and early Frank fighters.

Army Type 100 Gunsight with cushion protector at left. The knob is the dimmer switch for the filament for the reflected image. This sight became the most widely used Army gunsight of the Pacific War. *Courtesy of Todd Pederson*

Facing views of the Army Type 100 Gunsight. In addition to the reflected reticle, this gunsight also had a standby bead and circular cross hair sight at the side. *Courtesy of Todd Pederson*

Army Type 2 Mk.1 Gunsight *Ko*, front left view, built by Nippon Kogaku Kogyo K.K. Sights of this type were used on twin engine bombers, such as the Ki-49 Helen and Ki-67 Peggy. *80-G-192205*

Front right view. *80-G-192186*

Army Type 2 Mk.1 Gunsight *Ko*

This gunsight was designed as a flexible gunnery sight, and was found in general use on the Ki-49 Helen and Ki-67 Peggy. This was the only Army sight for turrets. It entered service in 1942. Knobs were provided that could pivot a ring and bead standby sight into the line of sight in case of lamp failure, and to pivot a tinted glass sun screen into place in front of the reflector plate when firing against bright clouds, since the reticle lamp only provided 10 candlepower. Two small knobs were also provided for making azimuth and elevation boresight adjustments. An unusual design feature was the two metal guards that could be raised when the sight was not in use to protect the vulnerable reflector plate from damage. The reticle image was identical to that of the Type 100 fighter gunsight. This was a small instrument, weighing less than a kilogram (1-1/2 lb). The body of the sight was painted the same neutral gray as Army cameras. Production for this sight was by *Nippon Kogaku Kogyo K.K.* (also written *Nikko*). This company was among the few that built equipment for both the Army and the Navy, since these two branches of the military were highly divided.

Right rear view. *80-G-192187*

Left side view. ***80-G-192185***

Army Type 3 Gunsight

A follow-on design and replacement for the Type 100 was the Army Type 3 Gunsight, officially adopted by the JAAF in 1944. Like the Navy Type 4 Gunsight, this sight was based on the German Revi 12C design removed from the purchased Junkers Ju 88. More compact than its predecessor, it weighed 1.36 kg (3 lbs). It employed a breakaway release mount for use in the event of a crash. This was an overall change from earlier designs. Quantities never exceeded that of the more widely used Type 100, yet the Type 3 was the last development of this type of Army gunsight that stayed in quantity production until the end of the war.

The most noticeable difference between the Type 100 and its successor (Type 3) was that the latter had a variable and a larger size reticle than preceding models, much like the Navy's Type 4 Gunsight Model 3. Among the aircraft using this newer Type 3 were the Ki-61 I and II Tonys, Ki-84 Frank, Ki-100, and presumably the Ki-45 Kai, used in the last stages of the war for intercepting B-29 Super Fortresses.

Various Other Army Gunsights

Other numbered gunsights were described in an Air Technical Intelligence Group Report No. 266—undated, yet apparently written soon after the end of the war because of the Japanese specialists that were interviewed. These were of an "Me" series.

The Me-1 was an all electric computing sight of which only two were built, both of which were destroyed in accidental aircraft crashes.

The Me-101 was also a computing sight for fixed guns consisting of a sight unit and two gyro units, one for azimuth and one for

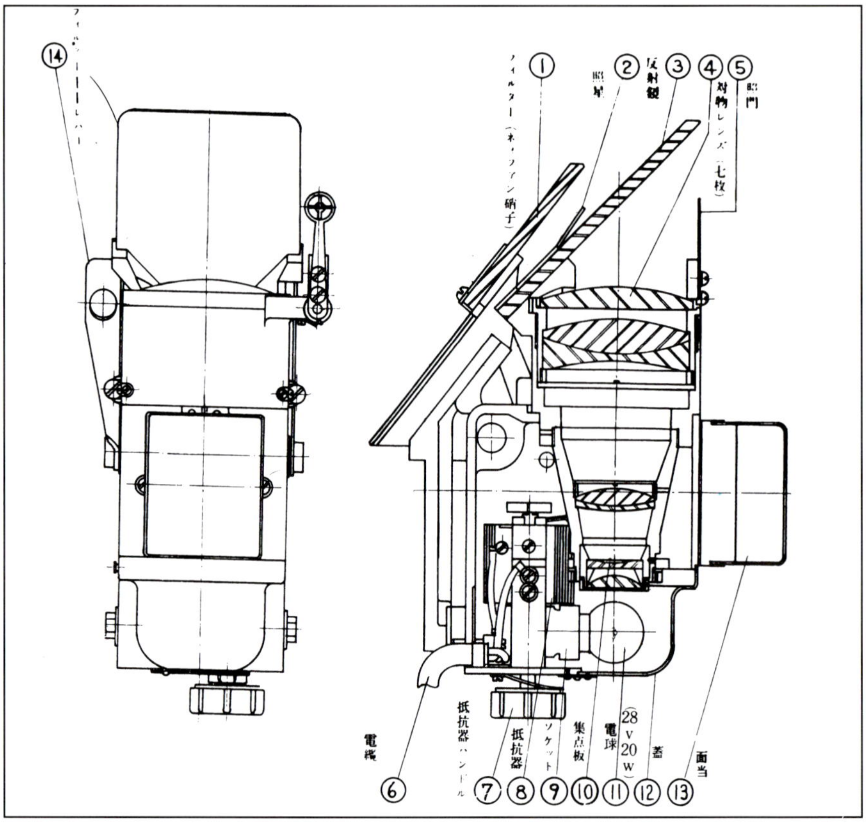

Army Type 3 Gunsight. This was the last mass produced gunsight for Japanese Army Fighters. This was based upon the Revi 12C German sight.
1. Filter (Neophan glass)
2. Auxiliary bead sight
3. Reflector plate
4. Objective (7 element)
5. Auxiliary right sight
6. Electric cord
7. Rheostat control knob
8. Rheostat
9. Lamp socket
10. condenser lens
11. Lamp (28v, 20w)
12. Lamp cover
13. Crash pad
14. Filter positioning lever

elevation. The war ended as testing was being completed. Although large and heavy, its performance was promising, but Japanese technicians felt this would compare favorably with the sights they found in the American P-47 and P-51 fighters.

The Me-103 was also a computing gunsight for fighters, having two main units, consisting of a sight head and a combination unit of computer and transmitter that contained two electrically driven spring restrained gyros. The coupling between the two units was pneumatic rather than electric. Tests were inconclusive by the time the war ended.

Bombsights

Level bombing tactics for the Japanese began with single engine aircraft, which required the bombardier to be located mid-point of the aircraft. Bombsights were therefore designed to be mounted on the floor and protrude through the outer skin of the aircraft when used for sighting. Even the Japanese Navy's first production twin engine bomber, the G3M Nell, had a solid nose, with the bombardier relying solely upon the optical view through the bombsight protruding beneath the aircraft. As other multiengine Army and Navy bombers came into service, the bombardier was eventually moved forward to occupy a glassed-in nose station.

Navy Bombsights

Navy Type 90 Bombsight

The Navy Type 90 Bombsight was an adaptation of the Goerz-Boykow model imported from Germany and Austria in 1927. This design was soon manufactured in Japan and became the Navy's standard optical bombsight of the Pacific War. Thirteen improvements were made during its service life. Designations for this bombsight began with the Type 90 Mk.1 and later developed into the Type 1 Mk.1 Bombsight, and finally as the Type 4 Mk.1 Bombsight.

The Type 90 Mk.1 Bombsight had a more rugged structure than the Goerz-Boykow Bombsight. These sights were manufactured by

Navy Type 90 Mk.1 Bombsight Model 1, manufactured by *Nippon Kogaku Kogyo K.K.* (Nikko), mounted in an upright operating position stand. This was a standard bombsight for the Navy. View looks forward at top. *80-G-192682*

This vertical view into the middle crew position of a Nakajima B5N Kate shows the Navy Type 90 Model 1 Bombsight in the operational position *80-G-194462*

Navy Type 90 Mk.1 Bombsight Model 1, showing inner components with the case cover removed. *80-G-169697*

Nippon Kogaku Kogyo K.K. (*Nikko*) from 1938 to August 1945. They weighed 14.0 kg (26.4 lb) and were 109-cm (3' 7") long, making them a large piece of optical equipment, especially in relatively cramped crew locations. Its optics could also serve as a drift sight. The internal optical sighting system employed a level bubble (rather than fixed crosshairs) as the sighting index, thus assuring a vertical reference line on which to base the sighting angle. The bubble appeared as a small circle that would wander around in the field of view as the sight was tipped, but in a manner consistent with the apparent motion of the terrain induced by angular perturbations of the telescope. Thus, the bubble would tend to stay aligned with the terrain image, except for the forward motion of the aircraft.

The Mark 1, or first version of the Type 90, was officially accepted by the Navy in 1930, as the Type number indicates. At the outbreak of the Pacific War its design was upgraded to the Type 90 Mk.1 Bombsight Model 2 Kai 5, and by now equipped all major land and carrier based attack bombers.

In operation, course corrections were orally transmitted from the bombardier to the pilot *via* gosport tube. Longitudinally, the principle of this sight relied on two coincidence points between the sight bubble and the target. Pre-mission planning produced a so-called C-value and an initial sight line depression, which were then set into the sight. During the bombing run the bombardier waited until the target appeared at the top of the field of view of his sight. When the target reached the center of the aiming bubble a timing clock was started that drove the rotating prism, and thus the sight line of the bombsight. To the bombardier it appeared that the target,

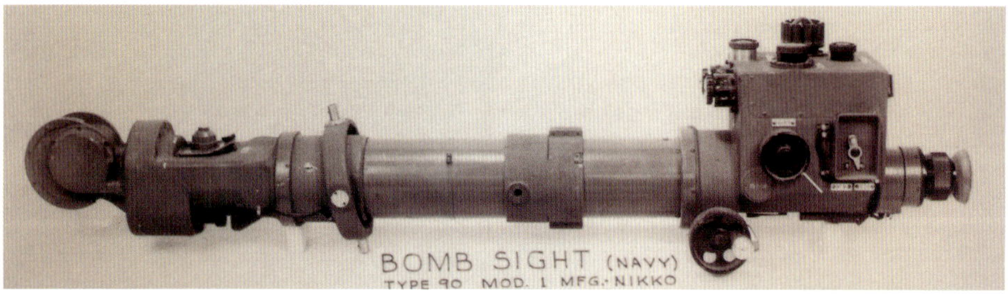

A closer view showing the internal workings of the Navy Type 90 Mk.1 Bombsight Model 1. Note the rate dials that would be exposed through the three openings seen on the top case cover. *80-G-169698*

Navy Type 90 Mk.1 Bombsight Model 1, which was operated in this vertical position. Note the gimble ring for supporting the sight about one-third the distance from the bottom. The wheel-like lense housing protrudes beneath the aircraft. *80-G-192683*

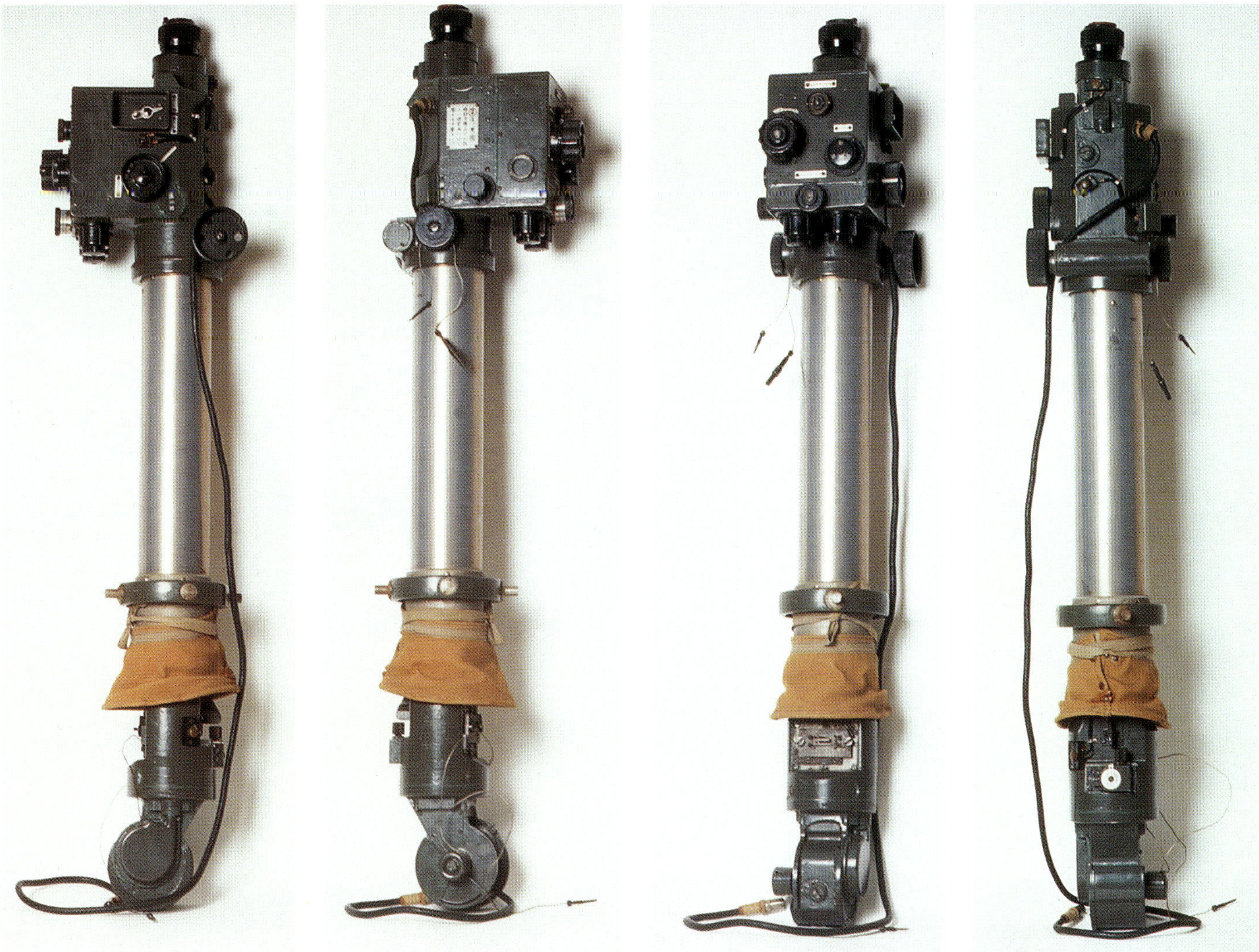

From left to right: Right side view; Left side view; Front view; and Rear view. Above and following *Courtesy of Todd Pederson*

Detail of sight head, left.

Detail of sight head, right.

Detail of sight head, front.

Operator's view of sight head.

Detail of sight head, rear.

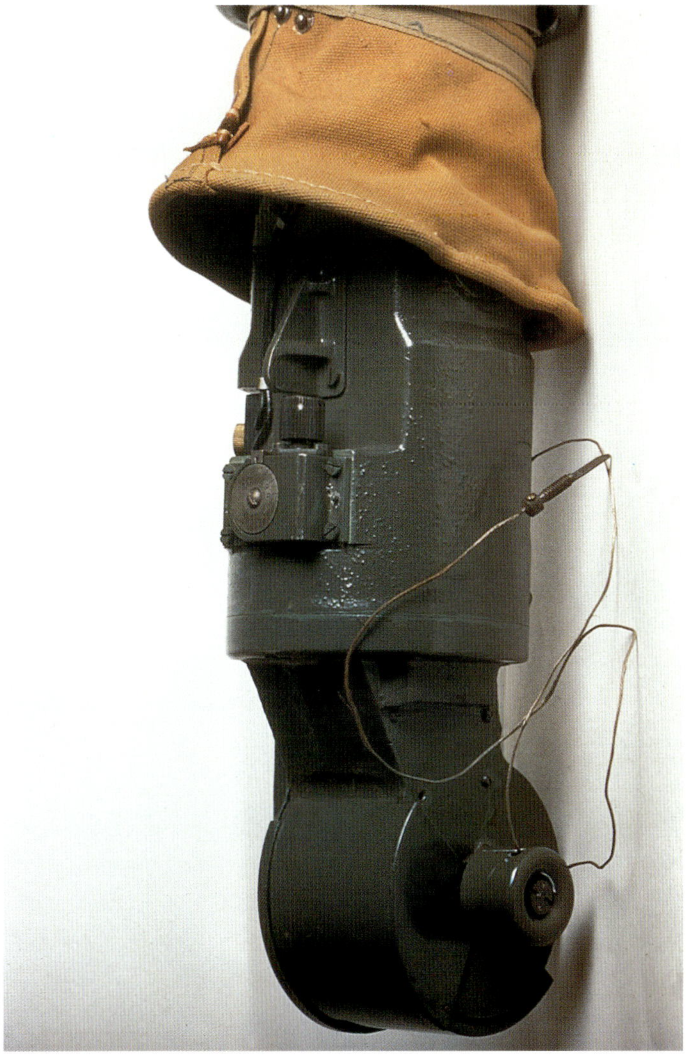

Sight head that protrudes beneath the aircraft is shown here. Note the canvas boot to cover the sight port when installed in the floor. Cables operated the sight head cover doors, protecting the lens when not in use.

Top view of sight control panel for Navy Type 90 Bombsight. The left placard points to the C-Value Indicator, while the one on the right points to the Trail Angle scale. The clock at the left records minutes and seconds of Drop Time, while the indicator at the left shows the drop angle.

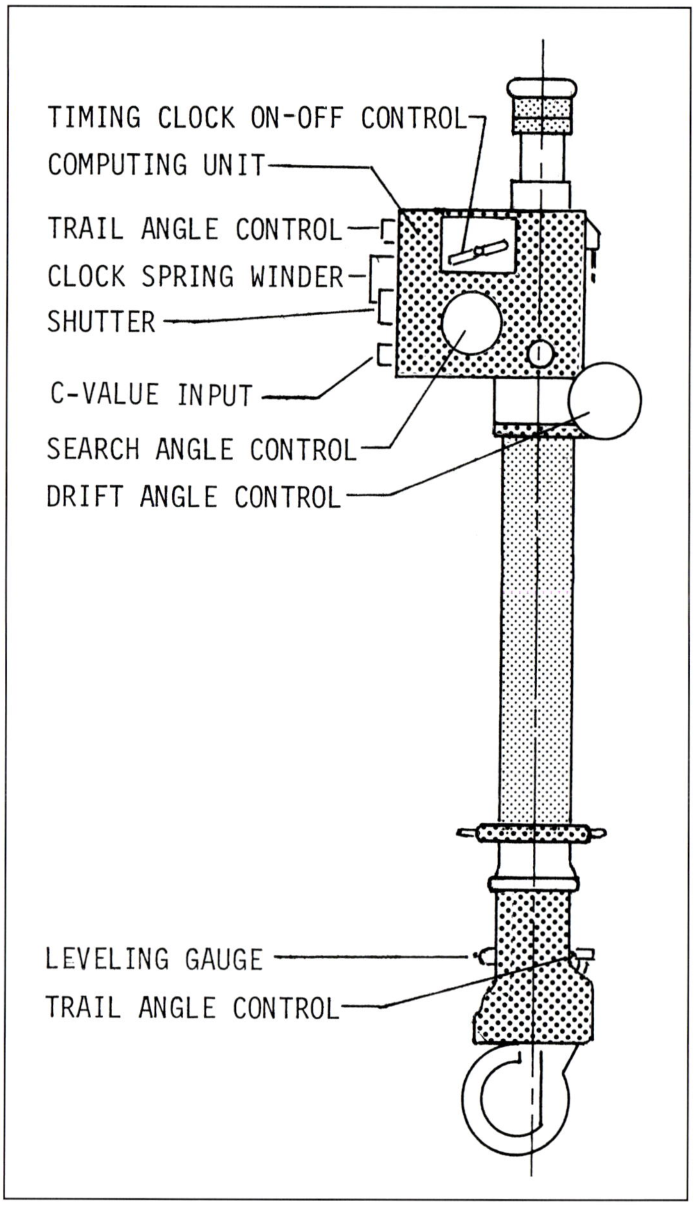

Above: View of sight optical prisms for forward angle vision. Right: Navy Type 90 Mk.1 Bombsight.

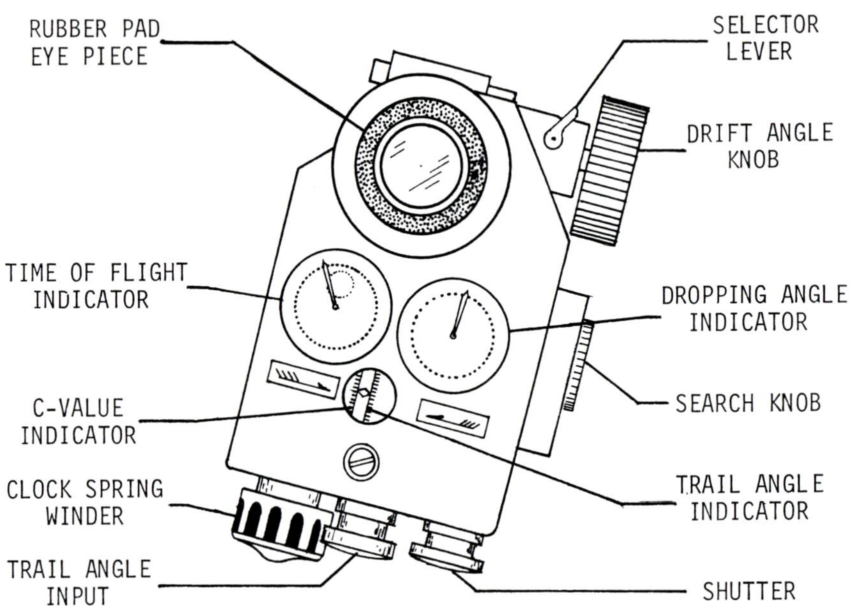

Control head components and optic view for Navy Type 90 Mk.1 Bombsight.

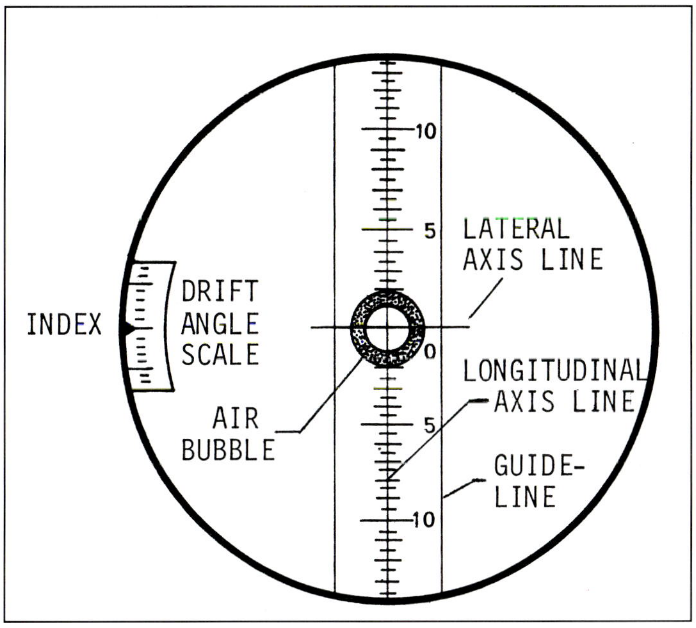

View finder image for Navy Type 90 Mk.1 Bombsight.

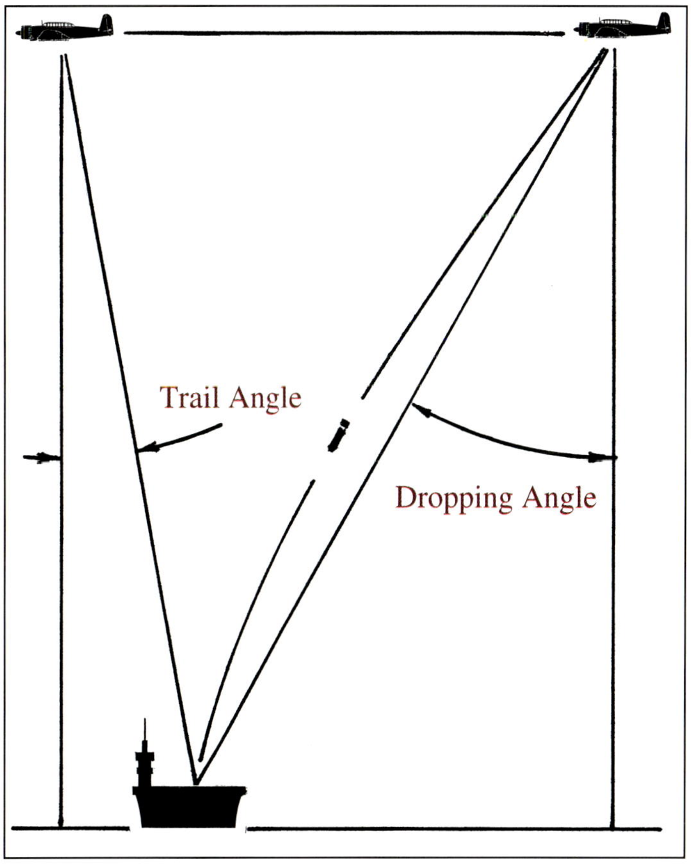

Level Bombing Triangulation.

which had been moving steadily down across his field of view until it reached the center, suddenly stopped and began to move slowly upwards again. This was because the sight line initially moved at a faster rate than the line of sight to the target. However, the relative motion slowed and finally reversed, due to the so-called "Boykow effect," and the target again appeared to descend towards the center of the sight. At the instant the target became centered in the bubble for the second time the bombs were dropped.

As can be imagined, the use of a bubble as an aiming index left much to be desired. The sight was greatly improved when the bubble was replaced by a Kugisho-type stabilizer developed by Lt. Cmdr. (Eng.) Furusawa.

At this late stage of the war the Japanese Army tried to manufacture the Navy Type 90 Bombsight, but the war had progressed too far for adequate production. Consequently, the Navy transferred considerable numbers of this sight to the Army for their operational use.

Navy Type 92 Bombsight

Allied forces were often stunned by Japanese bombing attacks when cloud ceilings were considered too low for level bombing. Instead of using optical sights that required greater altitude, a different type of sight (e.g. the Navy Type 92 Bombsight, using mechanical rather than optical sighting elements) was used. These were often mounted on the right side of the cockpit and operated by the pilot.

This Navy Type 92 (vector-type) Bombsight Model 2, which had no optics, was used effectively by the Japanese when ceilings were low and optical sights could not be used. Believed to be similar to RAF Mk.2 bombsight. Manufactured by *Tokyo Koku Keiki* K.K. *80-G-191669*

Anticipated speed and altitude were preset for the attack, and the bombs were released manually, often obtaining remarkable results under adverse conditions.

Navy Type 1 Bombsight

This sight was an improvement over the Type 90 in that stability was attained by adding a gyro stabilizer to the sight. This increased its weight by 2.5 kilograms, but bombing accuracy was also increased by one third. This sight became operational in April 1943 as the Navy's primary instruments. These were placed aboard Betty bombers in time for the air battles of the Solomons, where this sight was credited with a hit ratio of 32 percent at altitudes of 2,000 meters (6,562 ft.) and 26 percent at 4,000 meters (13,125 ft.).

In the meantime, the Japanese Navy was not satisfied with the Experimental Type 96 Bombsight that was of the synchronous type, nor by the Lotfe 7B purchased from Zeiss in Germany. As early as November 1939, Lt. Cmdr. Masami Kojima of the Naval Air Headquarters proposed large-scale tests to improve bombing accuracy. These were soon underway in three phases, each to confirm effectiveness of the Kugisho-type Stabilizer. It was through phase one tests that it became apparent that the use of autopilot would be necessary on bomb runs with high speed, large bombers. Additionally, it made improvements in the autopilot and bombsight gyro stabilizer that were so necessary for bombing.

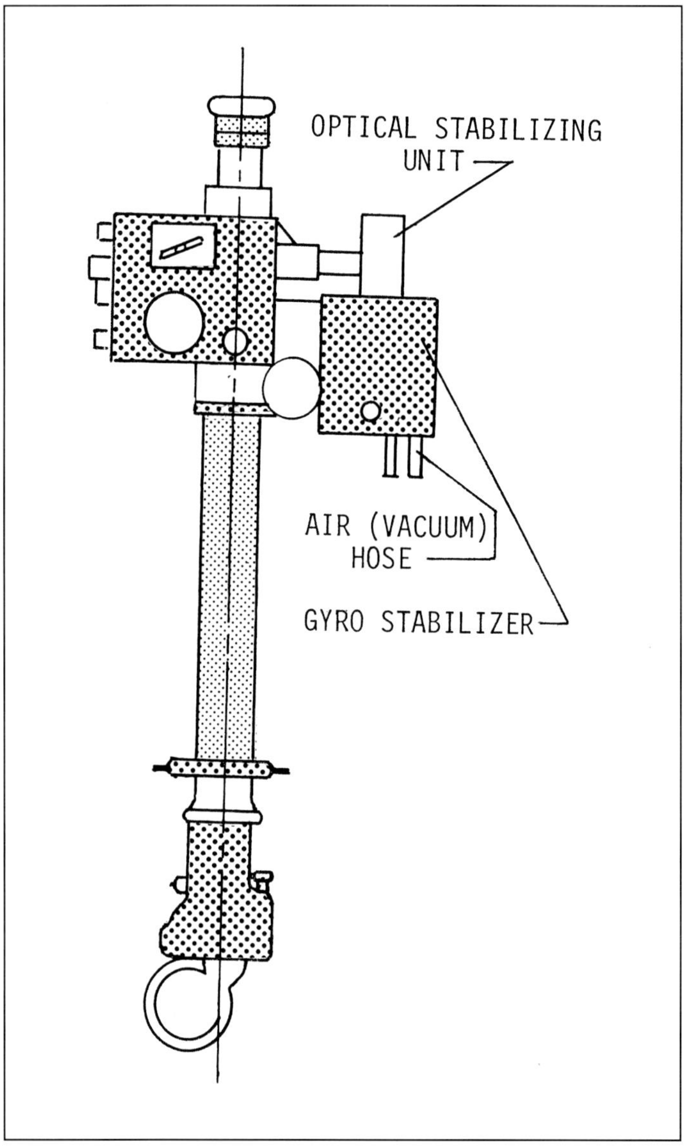

Navy Type 1 Mk.1 Bombsight.

In order to improve upon the Type 90 Bombsight itself, a meeting between Kugisho and *Nippon Kogaku Kogyo* officials was held in February 1941. After heated discussions for two hours, it was decided to develop a new bombsight according to the plan that Kugisho staff members proposed.

Prototype production began immediately, and the first was completed in December 1941 and accepted by the Navy as the Type 1 Bombsight. Phase Two of the program showed that the new design had an average impact error 2/3 that of the former. This meant that CEP (Circular Error Probability) would double with doubled altitude at 4000 m. The air driven gyro used for this test was not reliable, and therefore it was replaced by an electrically driven gyro.

The first models of this new type went into operational service in April 1943.

Navy Type 4 Mark 1 Bombsight

As the war situation deteriorated daily for the Japanese, air superiority turned in favor of Allied Forces. Japanese forces were retreating from Guadalcanal. The logical option for Japan was to turn to night bombing. In May 1943 it was decided to change from the existing telescopic bombsight to an all-new illuminated reflector type system for night operations. This became the Navy Type 4 Bombsight, which was developed within several months and placed in production in the spring of 1944. These were installed for the first time in a Japanese airplane in the nose of a Betty, much like the installation in U.S. aircraft with the Norden Bombsight. At night, a target could be recognized by eye from a 4,000 m altitude. Conversely, when using a telescopic sight, a target could only be recognized from an altitude of 2,000 m when under full moon conditions. For daylight operations the sight required a filter to reduce light intensity.

Navy Type 4 Mark 1 Compact Bombsight

In the summer of 1944, front-line *Kokutais* requested a smaller sight, one that was more suited for smaller carrier attack bombers such as Jill for low altitude level bombing between 50 and 100 meters. What evolved was the last operational bombsight developed in Japan.

In charge of this optical design at Kugisho was Lt. Cmdr. (Eng.) Nakajima, who completed the work in four days. Drawings for production were ordered and completed by Lt. (Eng.) Inoue and Lt. Cmdr. (Eng.) Takagi of Kugisho in ten days—a remarkable feat, pressed to completion by the urgency of the war effort.

After two months in construction prototypes were tested in a Nakajima B6N Jill, and the sight was placed in production immediately at the Ofuna Plant of *Tomioka Kogaku K.K.* The war ended just as mass-production was about to begin.

The Navy Type 4 Mk. 1 Compact Bombsight was designed for the smaller, more cramped conditions in carrier borne aircraft. *Courtesy of Todd Pederson*

This bottom view of the Navy Type 4 Mk.1 Compact Bombsight is one retained in the National Air and Space Museum collection.

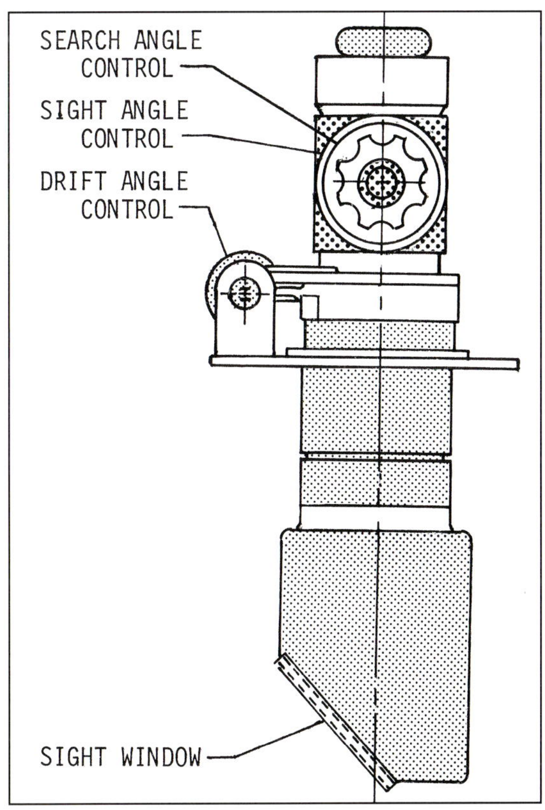

Navy Type 4 Mk.1 Compact Bombsight.

Navy 16 *Shi*, 17 *Shi*, and 18 *Shi* Experimental Bombsights

The 16 Shi Experimental Bombsight was the same as the Type 1 Bombsight, but installed within the airframe with a set of free standing stabilizing gimbals. With further development it became the 17 Shi Experimental Bombsight that used two sets of stabilizing gyros. Lateral angular offset of the bombsight sight line is automatically related to the longitudinal sighting angle through course correction, or drift angle. The two angular offsets, in concert, serve to bring the target image, as reflected off a stabilized mirror, into the bombsight's crosshairs. The principle of lateral course sighting correction control of the 17 Shi Experimental Bombsight Model 2 was more advanced than that of the Norden. Credit for this advanced design belongs to Capt. (Eng.) Shigehara Kitagawa, who was assigned to Kugisho. This sight was scheduled to be developed as the 18 Shi Experimental Bombsight in 1943 with change from telescopic sight to reflector sight. This was in parallel development at this time with the Navy Type 4 Bombsight that used the reflector sight concept. Prototype manufacturing of the 16 Shi was done by Nippon Kogaku, and the 18 Shi by Kugisho. The 18 Shi prototype was completed in December 1944, but the 16 Shi prototype was not completed until April 1945. Ground testing of both prototypes was completed, yet flight testing aboard the Rita bomber had not begun at the time the war ended.

Navy Type 91 Observation Telescope *Kai* 4

Before the Pacific War naval engagements were anticipated to be a war between surface ships. Smoke screens were part of the plan that would prevent the enemy from visual contact for laying gunfire. To counter this, an observation plane flying above the smoke could take angular measurements between the friendly and enemy ship so as to determine distance and azimuth, and radio this information to its mother ship.

The Navy Type 91 Observation Telescope was placed into service in 1931 and remained the primary instrument for this purpose throughout the war. This device looked very much like a drift sight, having a different grid designed for angular measurement. First used on such aircraft as the Nakajima E8N1 Dave, their later use was aboard the Mitsubishi F1M2 Pete, which was designed primarily for this fleet observation duty.

Navy Drift Sights

A predominant Type 97 designation applies to a number of Drift Sights with several configurations. These carried different model and *kai* numbers for their respective identities. Some of these were mounted horizontally when in use, protruding through an opening in the side of the fuselage, rather than extending through the floor. It is possible that some of these sights also served as Observation Telescopes for gun laying purposes when working with the fleet.

Navy Type 97 Drift Sight Mk.1, manufactured by *Nikko*. Used as a standard instrument in virtually all Japanese Navy aircraft, this one was removed by TAIU from a downed Aichi E16A1 Paul at Cancaabata Bay, November 4, 1944. Shown here mounted as it would be situated in an aircraft. *80-G-192684*

A closer view of the viewing head of this Navy Type 97 Drift Sight Mk.1. This sight appears as complicated as a Bombsight. There have been indications that a Bombsight was capable of performing the functions of a drift sight. *80-G-191292*

A three-point mounting bracket held this Navy Type 97 Drift Sight Model 4. *Courtesy of Richard Lane*

Another drift meter is this Type 97 Drift Sight Mk.1 Model 4 *Kai* 1. This was a more compact variant of the Type 97 Drift Sight Mk.1. Most likely, this instrument operated horizontally, protruding out the side of the aircraft. ***Courtesy of Todd Pederson***

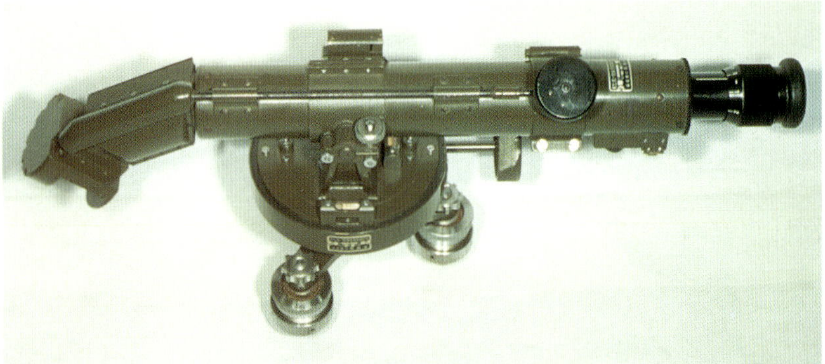

Type 97 Drift Sight Model 4 used by the Japanese Navy and now in a private collection and in original condition. ***Courtesy of Richard Lane***

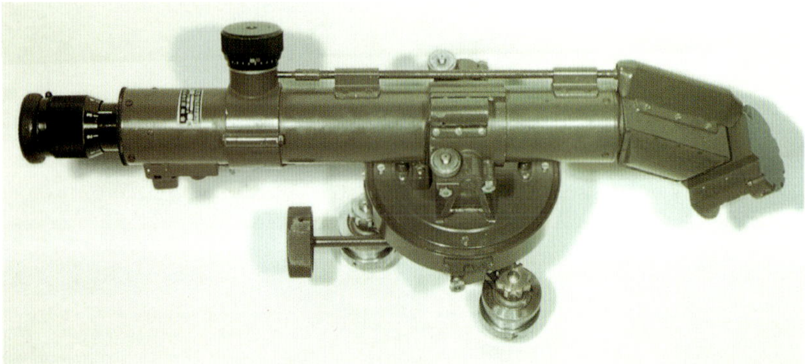

This view shows the opposite side of the Navy Type 97 Drift Sight Model 4. ***Courtesy of Richard Lane***

Eyepiece and adjustment knob for the Navy Type 97 Drift Sight Model 4. ***Coutesy of Richard Lane***

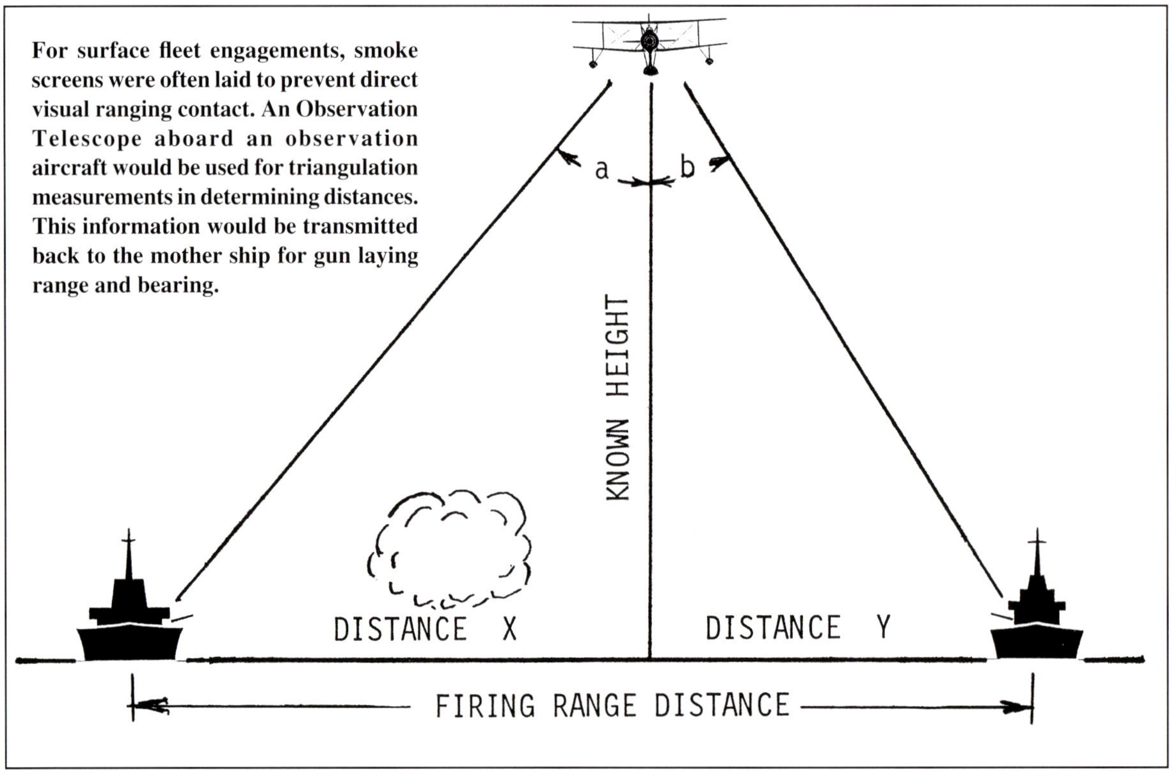

For surface fleet engagements, smoke screens were often laid to prevent direct visual ranging contact. An Observation Telescope aboard an observation aircraft would be used for triangulation measurements in determining distances. This information would be transmitted back to the mother ship for gun laying range and bearing.

This is a Navy Type 97 Drift Sight typically found in many Japanese Navy aircraft. This one was recovered from a Paul in November 1944. Interior photos of NASM's *Seiran* shows this instrument installed. *80-G-191291*

Army Bombsights

The first bombsights used by the Imperial Japanese Army were taken as spoils of war from Germany following World War I. These were made by Zeiss and Goerz. Neither of these sights had drift correction capability. After evaluating the two bombsights, the Japanese Army contracted with *Nippon Kogaku Kogyo K.K.* to copy and manufacture the Goerz design. This became the Type 13 Bombsight.

Army Type 88 Bombsight

By 1927, the Army realized the necessity of drift angle correction, and therefore had the manufacturer design this capability into an improved version of the Goerz sight. When accepted as official equipment, this became the Army Type 88 Bombsight. In addition to *Nippon Kogaku Kogyo* (*Nikko*), production was shared with *Tokyo Kogaku Kikai K.K.* (*Toko*).

Army Type 99 Light Reflector Bombsight

Bombsight design concepts took on a major change with the introduction of the Type 99. Instead of the often awkward to handle floor mounted vertical shaft bombsight, this reflector type was less complicated, lighter in weight, and easier to handle. Much of this design change was based upon having multi engine bombers, whereby the bombardier was placed in the nose of the aircraft for a broader range of view. These bombsights were better suited for low level bombing when needing a greater forward view for target acquisition.

Army Type 88 Bombsight manufactured by *Nikko* and *Toko*, mounted in an upright operating position stand and photographed by TAIU as it would be in an aircraft. The box unit would be against the operator. This was a standard Bombsight for the Army, used in Sally, Lily, Helen, Peggy, and others. *80-G-192685*

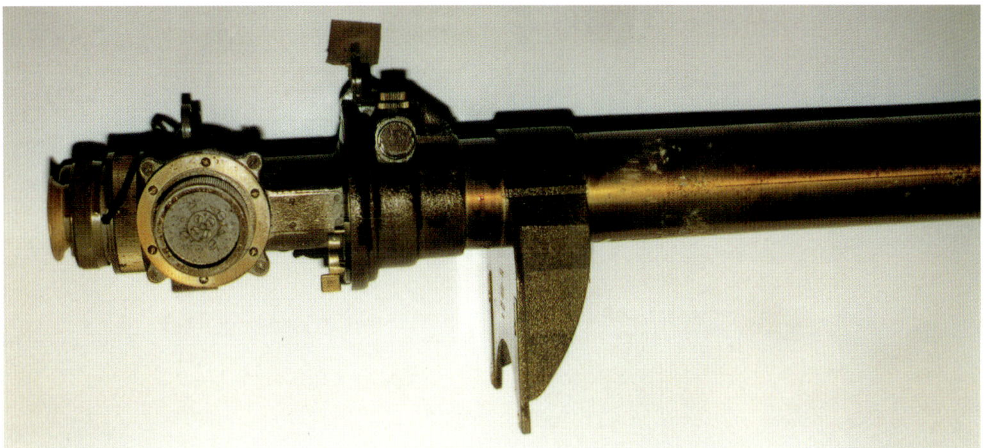

Army Type 88 Bombsight Model 4 in the NASM collection coated with preservative. On the bracket facing the operator was placed one of several types of mechanical calculators, most likely used for computing drop angle and wind drift.

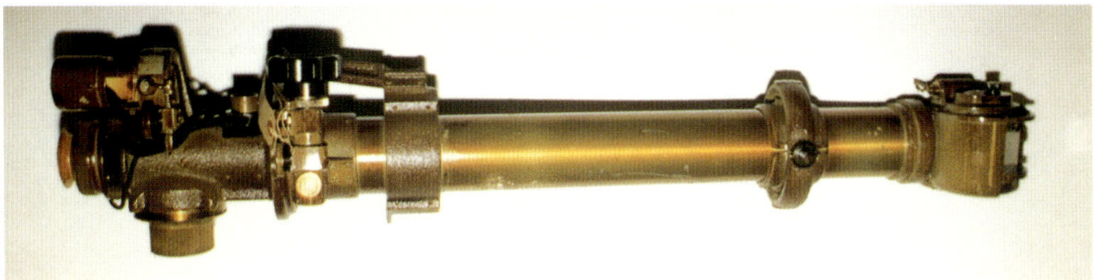

Front view of the Army Type 88 Bombsight Model 4 that faces away from the operator. The midpoint collar is half of a gimble mount that is in addition to the round opening found in the floor of bombers through which the lower portion protrudes.

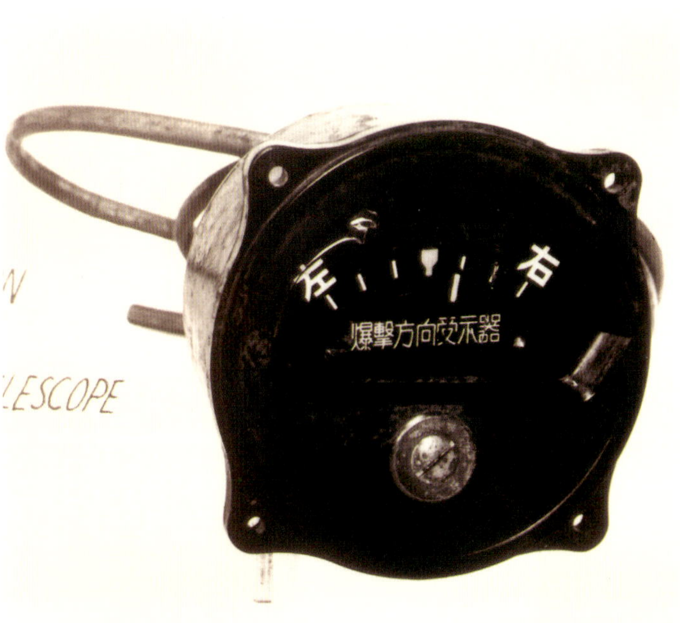

Part of the Army Type 88 bombsight system is this PDI (Pilot's Direction Indicator). Actual translation is "Bombing Direction Receiver." These are found in the pilot's instrument panel and give left-right corrections from the bombardier when on the final bombing run. *80-G-192543*

Top right side of the Army Type 88 Bombsight optical and computing head.

This lower end of the Army Type 88 Bombsight protrudes into the airstream, forward is to the left. An optic window is on the curved portion. The supporting gimble above rests in a collar that provides power to the unit and transmits directional changes to the pilot.

The Army Type 88 Bombsight came with two of these Bomb Release Table Boxes, probably calibrated for different conditions. This was secured to the bracket on the sight. Mechanical adjustments moved pointers over selected card-graphs that were inserted under the window.

Army Type 99 Reflector Bombsight. This model would have been found on Sally, Lily, Helen, and Peggy as a replacement for the Army Type 88 Bombsight. ***Courtesy of Richard Lane***

Army Type 99 Reflector Bombsight manufactured by *Nippon Kogaku*, and contained in a private collection. *Courtesy of Richard Lane*

This Army Type 99 Reflector Bombsight was one of several taken by TAIU for evaluation. This shows the unusual left side mounting base. Similar views in the TAIC collection are 80-G-192562 and '563. *Courtesy of the U.S. Air Force Museum*

Identified in TAIU photos of captured Japanese equipment, this bombsight is an Army Model 2, built by *Tokyo Kogaku Kikai K.K.* These would have served the same purpose as the Navy's vector type bombsights for low level bombing. This model appears to be more advanced in design. *80-G-192559*

Army Type 1 Bombsight

Improvements continued on the Type 88 until 1941, when official acceptance was made by the Army of what was then the Type 1 Bombsight. This vertical design concept remained more practical for high level bombing.

Army Model 2 Vector Bombsight

Perhaps sparked by the Navy's success with using vector type bombsights in low level bombing attacks, the Army also adopted this technique and its related equipment. Twin engine bombers brought this technique to the forefront in that the bombardier, now in the nose, had a further forward view of the target, and was thus able to have better alignment for a longer bomb run.

Army Model 10 Bombsight

From a fallen B-24 Liberator over Rabaul in 1942, the Japanese military was able to copy the highly secret Norden Bombsight, producing separate bombsights for both services using the same principles as the Norden. For the Army, this became the Model 10 Bombsight. By the end of the Pacific War the Type 1 and Model 10 Bombsights were standard equipment for the Army, although the latter had not reached quantity production.

Very similar in appearance to the Norden Bombsight is this Army Model 10 Bombsight. From American instruments captured on Rabaul, this highly secret bombsight was revealed to the Japanese.

This working front view of the Army Model 10 Bombsight reveals the gimble mounted gyro inside the large opening. Few of these were used in combat before the war ended.

Aircraft assignment of gunsights and bombsights

Navy Gunsights	
Type 95 Telescopic Gun/Bombsight	Claude, Pete, Val 11
Type 98 Reflector Gunsight Model 1	Zeke 21, 22, 32, 52, Rufe, Rex 11, George 11, Jack, Paul
Type 98 Reflector Gunsight Model 2	Judy 11, 43, Irving, Paul
Type 99 Telescopic Gun/Bombsight	Val 11 & 22, Pete, Claude
Type 2 Telescopic Gun/Bombsight	Val 22, Judy 12, 33, *Seiran*
Type 3 Reflector Gun/Bombsight	Frances, Grace, *Seiran*
Type 3 Compact Gunsight Model 1	Night fighters Irving, Frances, Myrt, Judy
Type 4 Gunsight	Zeke 62, George 21, Jack 21, 31, 33, *Shinden*
Type 4 Compact Gunsight	Betty 22, 24, 34, Frances, Emily, Rita
Model 2 Reflector Gunsight Mk.1 *Ko*	Most flexible gun positions
18-Shi Experimental Gunsight	Rita
19-Shi Experimental Compact Gunsight	Night Fighter Jack

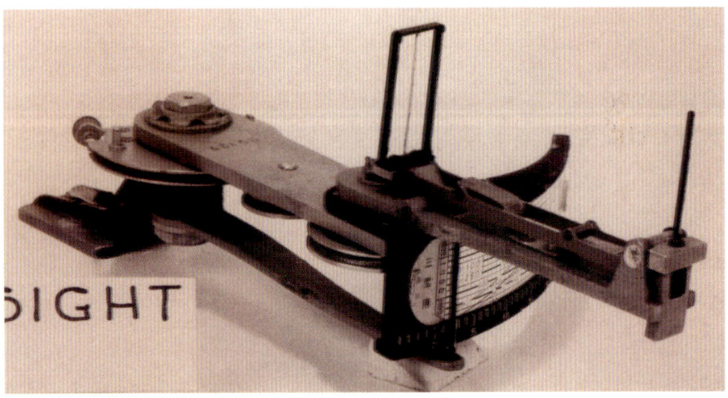

A number of items of interior Japanese aircraft equipment are contained in the TAIU photo collection contained in the National Archives. Among these photos are several Navy Torpedo sights like this one of unspecified Type and Model. ***80-G-192680***

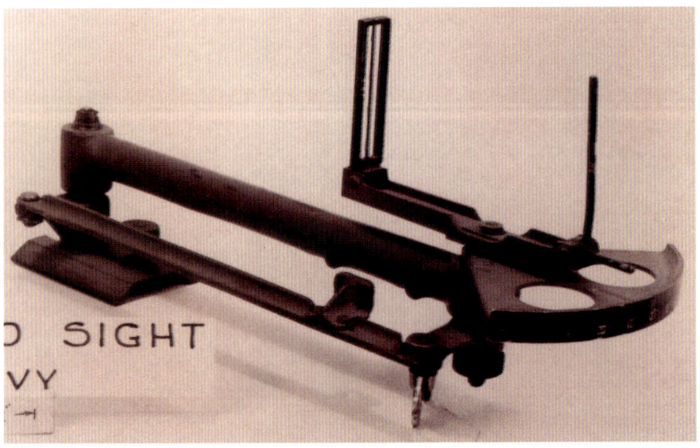

This Navy Torpedo sight of unspecified Type and Model. Off set angles and calculated speed of the surface craft, along with other variables, would be set into these sights for torpedo launch alignment. ***80-G-192677***

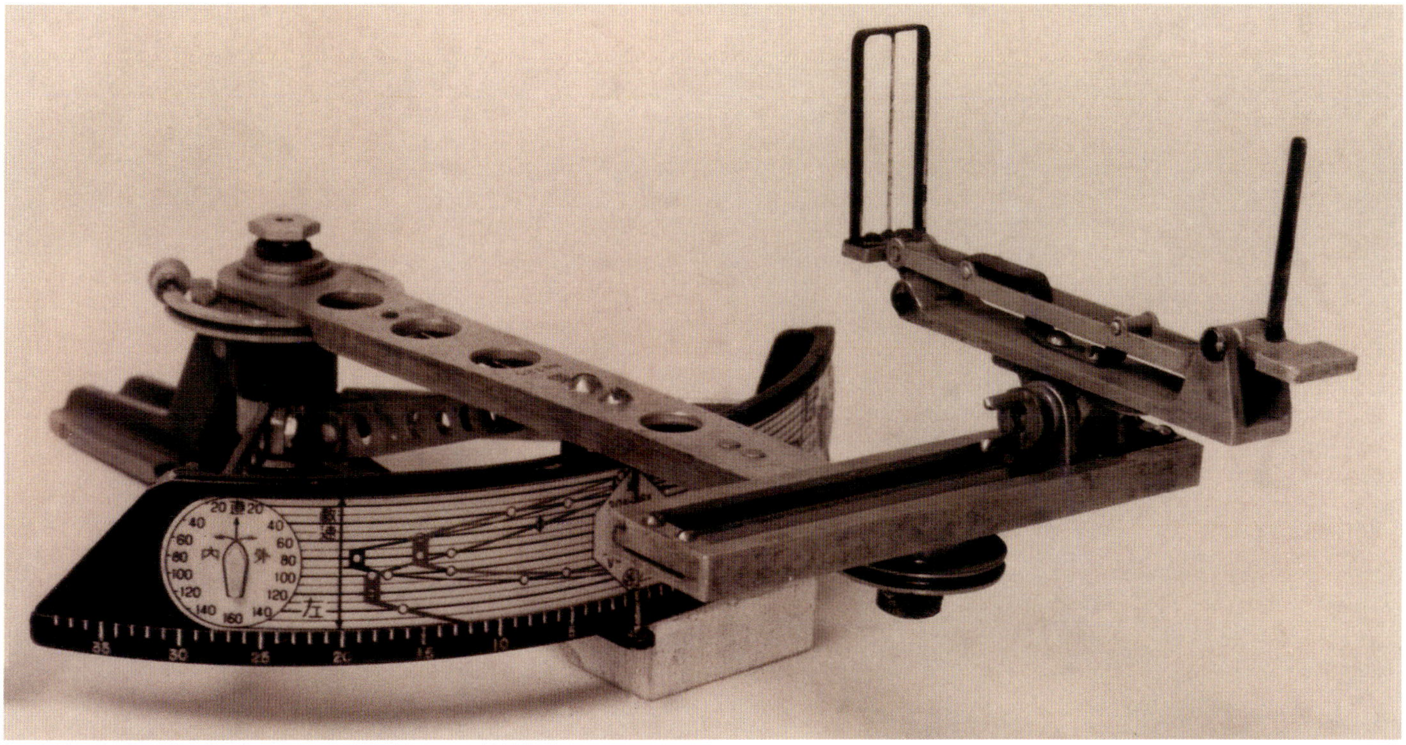

Navy Torpedo sight of unspecified Type and Model. Being of large format, this is a type that would be found on the larger multi-engined bombers, most of which were capable of torpedo launching. ***80-G-192567***

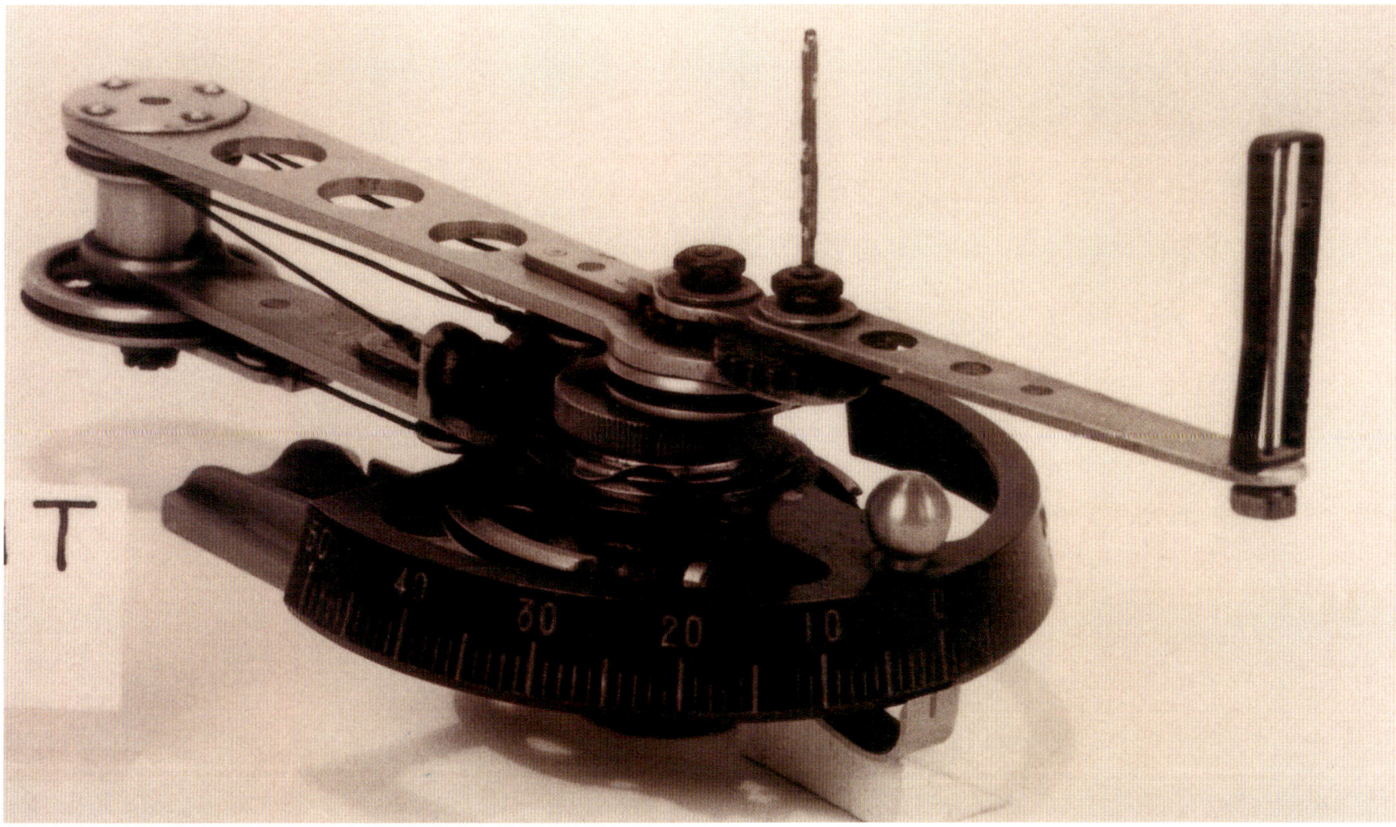

Navy Torpedo sight of unspecified Type and Model, but believed to be Type 97, and an instrument that might have been found in a B5N2 Kate torpedo bomber. ***80-G-192668***

Army Gunsights

Type 89 Sighting Telescope	Nate, Oscar 1, Sonia, Tojo 1, Ida, Ki-79
Type 100 Gunsight	Oscar 2, 3, Tojo 2, Tony 1, Nick
Type 2 Mk.1 Gunsight *Ko*	Sally, Helen, Peggy
Type 3 Gunsight	Oscar 3, Tony 2, Frank, Ki-100, Randy, armed Dinah 3

Navy Bombsights

Type 90 Bombsight	Judy, Kate, Jill, Grace, Jake, Betty, Frances, Emily
Type 92 Bombsight	Betty, Frances, Jill
Type 1 Bombsight	Betty, Frances, Jill
Type 4 Mark 1 Bombsight	Betty, Frances, Jill
Type 4 Mark 1 Compact Bombsight	Plan only, Grace, Jill
16 Shi Bombsight	Experimental
17 Shi Bombsight	Experimental
18 Shi Bombsight	Experimental
19 Shi Bombsight	Experimental

Other Navy Sights

Type 91 Observation Telescope	Dave, Pete, Alf, Jake
Type 97 Drift Sight	Many aircraft

Army Bombsights

Type 88 Bombsight	Lily, Sally, Helen, Peggy
Type 99 Light Reflector Bombsight	Lily, Sally, Helen, Peggy
Type 1 Bombsight	Lily, Sally, Helen, Peggy
Model 2 Vector Bombsight	Lily, Sally, Helen, Peggy
Model 10 Bombsight	Patsy

Notes:

[1] This may have been designated Type 4 Mk.1 Gun/Bombsight Model 3.

[2] Newly established in November 1943 for joint usage by Army and Navy.

7

Interior Colors and Coatings

Ideally, Japanese Navy aircraft interiors would have been painted one standard color, and Japanese Army aircraft interiors painted another. Such is not the case, however. Variations have become apparent not only within each service, but also by manufacturer, their aircraft, time periods, and aircraft equipment. For that reason these variations have been recorded here based upon examples studied so that readers can form their own conclusions as to any standards they believe may have prevailed. More important than having a sample from which to record the color, consideration must be given to the color shift that could have taken place over the decades since the paint was applied. Over this period, there may well have been color changes caused by light, moisture, and oxidization.

Fortunately, many of the interior paints are shades of green, and green is one of the most stable colors, red being one of the least stable. Where possible, electronic colorimeter readings were made of these samples of color and are included here for the record. Because of the thinness of some of these paint coatings, a colorimeter reading is good only to record what was found the day of examination and to provide a baseline for comparing future readings. In the case of airplanes having thin or severely marred coatings, a color chip with eyesight comparison remains the best method of recording these colors. To duplicate them with fresh paint is impossible, because aged paint, well oxidized and worn, cannot be taken from a can. Therefore, whatever coating is prepared for trying to duplicate an existing color must be regarded as a simulation, and the exact numeric readings must be that of colorimeter values.

Having said that, where then does this lead us in conveying color to the reader? Colors described in this chapter represent as closely as possible the aircraft interior and component colors, and can only be referred to as being *approximations*. It is not practical to describe samples of each shade examined, nor can we assume that the Japanese had that many colors in inventories. This author has summarized most of these color samples by grouping them into *color families*, meaning those colors that may well, at one time,

This is the unrestored nose section of the Mitsubishi G4M3 Betty in the National Air and Space Museum. It retains hard evidence of multiple shades of green throughout the interior.

This before and after comparison of the Kawanishi N1K2-Ja cockpit shows a duplication of the original colors that have all but deteriorated away. To have merely preserved the interior would not be representataive of its operational appearance.

have been intended to be the same color. Therefore, these color families of small variation are represented by one color value that is a close average.

In the two books by Donald Thorpe, *Japanese Army Aircraft Colors and Markings of WW II* and *Navy Aircraft Colors and Markings of WW II*, a fairly well standardized and accepted color coding system has been established. Rather than to begin another system, it seems best to remain with the color coding of "A" for Army colors and "N" for Navy colors, A/N for colors found in both services, and to use the next 10 digits in that system for these interior and component colors, beginning with 31. Where colors have already been identified with an established code number, these codes are used.

It cannot be overemphasized that colors being discussed in this book cannot be regarded as "official," much less "original." Time has had its effect on color. What is stated here is an approximation as close as possible through visual estimation. It is up to the user of this information to interpolate the deterioration process when matching old colors to new color chips.

Although as many samples as possible have been examined and have become a part of this study, this is insufficient evidence to make broad statements. Where no data on colors for a certain aircraft exists, perhaps enough evidence pertaining to the same manufacturer will give the user sufficient guidance upon which to make constructive assumptions.

The Japanese language instruction markings add the finishing touch to this otherwise sterile appearing cockpit. Careful reconstruction of these placards must be finished before the evidence of their original existence is lost due to repainting.

This cockpit view of an Irving taken before its restoration shows a number of shades of green that were used. Many, however, are of the same family of greens and could have been intended to have been the same, but from different manufacturers.

Japanese Aircraft Interior Colors (Close Approximations Only)

The color values expressed in this section have been recorded as faithfully as possible by the best color sample and recording devices found in various color sample systems. No one system has adequately represented all the colors listed here as having been examined. For example, to have used the Federal Standard system throughout, and stating the actual color to be a little darker than, or a little more yellow than, adds a guess-work factor, and the value of accuracy is then lost.

The most rewarding source for Japanese interior colors has been the aircraft in the National Air and Space Museum, Washington, DC, followed by samples of materials from private collections. The listing of aircraft types given here does not imply a general overall interior coating of the color recorded, but contain major areas and components from which readings were obtained.

Colors were recorded and matched as closely as possible during the NASM restoration of this Irving night fighter. A near total repainting was called for, however, because of excessive corrosion of many of the working parts.

Record of Existing Original Japanese Colors

Samples found in:	Code	Closest Standard	Color
Ki-45Kai Nick * C6N1 Myrt Ohka 22 E13A1 Jake	N.1	Munsell 10GY 2/2	Black Green
B7A1 Grace J1N1 Irving Kikka C6N1 Myrt	N.2	Munsell 10G 3/2	Dark Green
Ki-100	A.31	Pantone 5743C	Kawasaki Green
D4Y Judy Ki-48 Lily Ki-49 Helen	A/N.32	FS 595b 34094	Interior Green
C6N1 Myrt J7W1 Shinden N1K1 Rex J1N1 Irving P1Y1 Frances A6M2 Zeke (Nakajima)	N.33	FS 595b 34082	Olive Green
Ki-46 Dinah N1K2-Ja George A6M Zeke 32, 22, 52 P1Y1 Frances G4M Betty	A/N.34	FS 595b 34127	Mitsubishi Green
Ki-100 (electrical comp.)	A.35	Pantone 462 U	Grey Brown
Ki-100 (cockpit floor)	A.36	Pantone 405 C	Dark Brown
Ki-45Kai Ki-61 Tony Ki-48 Lily	A.37	FS 595b 30118	Light Brown
A6M3 Zeke 32, 22	N.38	3.2Y 7.6/3.5	Sand

*Strangely, this Army aircraft (Nick) had a close match to a Navy color.

Aotake—the mystery coating

The mere mention of the word "Aotake" among two or more people that are familiar with the word and material is bound to produce differences of opinion. There is more mystery surrounding the paint material *aotake* than any other coating on Japanese aircraft. The problem is not only based upon trying to identify and catalog its color, but also on the question of its purpose. Its application throughout aircraft structures, both Army and Navy, is quite inconsistent. Aotake is not a color, but a coating. In many cases, aotake appears to have been applied to surfaces well before the parts were attached to the airframe, or even cut to shape. Often, interior inspection of an airframe reveals a quilt-like appearance; some pieces prepainted, others not. This has led to one belief that small subcontractors used aotake to mark a pattern to be cut or drilled, much like the purple ink that is used for that purpose today. This is generally on the inside of the skin. If aotake were a primer, it would have been applied on (outside) surfaces that were intended to have been painted.

The name "*aotake*" is also controversial. When translated directly from Japanese, the word "aotake" seems to mean "blue bamboo." However, "ao" also means "green" or "young" in Japanese. In this case, therefore, "green bamboo" or "young bamboo" may be more appropriate as translations of "aotake." For example, Japanese people call the green light of a traffic signal "ao-shingou" (blue signal), but the actual color of the light is green. For another interesting example, in Japan the expression "your bottom is still blue" means "you are still too young." This is because Japanese babies have a Mongolian blue spot on their bottoms. Following this reasoning, the color "aotake" can be described as the same color as young bamboo, which is usually green-yellow.

Whether to use *aotake* or *aodake* as the name of this material also raises controversy. Current dictionaries in Japan favor *aodake*, but for its usage during the Pacific War, users were accustomed to *aotake* and its usage exists to this day. Therefore, for the sake of this established recognition, this coverage will call it *aotake*. But now we are placing too much emphasis upon the readings of the kanji.

The simulation of an aotake coating becomes a tricky operation. In this case of duplicating the interior coating of the Irving night fighter, clear lacquer with a small portion of green was sprayed onto the surface until the desired appearance in color intensity was achieved.

In the case of restoring the cockpit of the Aichi M6A1 *Seiran* at NASM, paint was in good enough condition that a thorough cleaning and a wax coating preserved the original appearance of the cockpit, retaining its used look.

The Kawasaki Ki-45 kai Nick is an unrestored aircraft in the NASM collection. This Army cockpit shows a light brown A37 interior coloring throughout. Instrument panel is black that appears to have been a semigloss originally.

Nick was designed as a twin-engined fighter, but in the closing months of the war many were modified with two oblique mounted cannon for use as a night fighter.

This is another unrestored aircraft in the NASM collection. This Kugisho P1Y1Frances was Japan's most advance bomber placed into operational service by the Navy. When first encountered it was thought to be a fighter because of its speed and maneuverability.

A variety of shades of green are shown in this view of the unrestored cockpit of Frances. This indicates that a number of subcontractors were used to equip this cockpit.

This photo of the unrestored back seat of Frances shows, among other things, the Direction Finder antenna and receiver control box at the bottom center of this picture. This unit is for the Navy Type 0 Ku Mk. 4 RDF. The large number of structural wooden substitution pieces is apparent in this view.

Encounters with aotake in protected areas, such as under tightly riveted pieces of metal, reveal a deep transparent blue. In areas where long exposure has taken place, aotake is much lighter and greener. When trying to match a color sample for record, almost every variance in lighting produces a different shade and color. Aotake seems to vary with every sample examined. Even on a small piece of metal, sometimes many colors can be seen as light conditions vary. They are blended together like oils on a canvas. Therefore, when simulating it for whatever purpose, *if it looks right, it is right.*

The Kyushu J7W1 *Shinden* is a fascinating example of Japan's futuristic fighter that was being tested at the end of the war. This airplane became an example of Japan's most advanced designs and installed equipment. It resides unrestored in the National Air and Space Museum collection.

Simulations of this paint for aircraft restorations generally start with clear lacquer, and contain small amounts of medium green. They are mixed until the desired intensity of green is reached, making sure that the material remains translucent. Translucence is also governed by the amount of material applied to the surface. *The tendency is to overstate the green.*

Mr. Mitsutoshi Okano, Chief, Historical Material Office, of Mitsubishi Heavy Industry Co., Ltd. in Nagoya, Japan, supports the idea of "if it looks right, it is right," because there was no standard color sample available in Japan during WWII. Also, Mr. Okano could not locate anything from Mitsubishi's historical files, and believes such information might have been burned during the war or destroyed by the famous Isewan Typhoon. So, when Mitsubishi restored their A6M5a (s/n 4708), now located at Mitsubishi's Nagoya Aerospace System Works, they used a similar color paint based upon what was found in actual Zero airframes. This paint is "Hi Urethane, #5000 Aotakeiro, MDIC-4352" made by L. Mizuo Co., Ltd.

The coatings of paint found in the one surviving *Shinden* raise more questions than answers. The airplane was converted upon reaching the United States with the purpose of being flown, but this did not take place. What changes in coloring are possibly American, while others are Japanese, leave much room for debate. Note the improved cockpit design of less exposed structural members.

How aotake was applied and over which parts varied by airplane type, and became somewhat inconsistent through the time span of the manufacture of certain aircraft types.

Japan entered the jet age with the first flight of this Nakajima *Kikka* on 7 August 1945. Shown here is Lt Cmdr Susumu Takaoka climbing in for the second but aborted takeoff of Kikka. One Kikka and parts of another reside in the National Air and Space Museum awaiting restoration.

The answers regarding this material were thought to be possibly found in Japan. As early as the 1960s, while stationed in Japan, this writer made an extensive inquiry into the questions about aotake. People associated with aviation during the Pacific War were aware of the material, but no one had any satisfactory answers. Replies were as many as the speculations already expressed here.

Someone in the United States in the 1970s felt they had seen this paint on the inside of a Datsun (now Nissan) car. If still being used, an answer could be obtained. Inquiries to Datsun and to Japanese automobile dealerships for a look under the hood and inside fenders revealed nothing. Seemingly, the use of aotake had disappeared. No one could give a positive answer.

So, what is it about aotake that makes it so different from other paints? Conservators at the Australian War Memorial in Canberra asked this question of themselves and found that, other than its translucence, it is not really very different from other paints of the era.

The AWM decided to carry out a partial laboratory analysis of the aotake found on a Mitsubishi-built Zero Model 21 (s/n 5783) they were restoring. Their purpose was, through the identification of the components of the substance, to determine why aotake was used. An interesting report dated 16 April 1984 was prepared by

For as advanced a jet aircraft as the *Kikka* was to Japan, this cockpit interior view has a crude appearance in design and fabrication. Note the many components that were painted before being assembled with the un-painted rivets. Simplicity of jet aircraft operation over that of reciprocating engines is apparent by the lack of valves, switches, and controls seen in other aircraft cockpits.

Chemist Bruce Ford, but unfortunately he became ill and could not complete his experiments. Rather than reproduce his report in this book, the author decided to provide the nontechnical summary, which follows:

Most of the inside skin of the Zero is painted with a blue-green pigmented lacquer. The lacquer is possibly, but by no means certainly a modified phenolic, and will be the subject of further investigation.

Mr Ford followed a basic and accepted approach for identification of the chemicals involved in the makeup of aotake. Since aotake was likely composed of several chemicals, the trick would be to separate them. Mr. Ford followed a published approach for paint using the solubility test contained in Hummel & Scholl: Atlas of Polymer and Plastic Analysis Vol. 3, 1981, Verlag. The range of chemicals used for solubility generally followed the rule of thumb—such as dissolves like—in other words, organic dissolves organic and inorganic dissolves inorganic. Chemical testing by Mr. Ford found that part of the paint dissolved in sulfuric acid, which indicated from the Hummel & Scholl reference that it must be a phthalocyanine. The dissolved chemical was next subjected to an infrared spectrum analysis, the result of which fully correlated with published data for Phthalocyanine, Pigment Blue 15.

What is Phthalocyanine?

A common dictionary revealed that phthalocyanine is a blue-green organic compound. A chemical dictionary mentions four commercially important forms:

(1) a water-soluble green, (2) a chlorinated copper green, (3) a copper version that results in Pigment Blue 15, and (4) a metal-free version having a blue-green color.

Phthalocyanine is used world wide to this day, and is a fairly common pigment. It is important because it is relatively cheap and stable. Pigments usually do not have any unusual properties besides pigmentation, so if there were some special characteristic of phthalocyanine, it would be in the published literature. A literature search revealed no such special characteristic. But in the 1930s and 1940s, paint technology was not as advanced as it is today, so the color variations must be due to some other influence, as generally the pigment will not color shift by itself.

So, if phthalocyanine has no special properties (characteristics) and is even used today for its stability as a pigment, it is therefore difficult to understand why the reported color changes occurred on Japanese aircraft of the 1930s era, and why almost every variance in lighting produces a different intensity of shades of color.

Mr Clive H Hara, chemistry consultant and President, Clive H Hara Inc., a chemical analysis firm, offers this information. Copper phthalocyanine pigments are and were available in two crystalline forms: the alpha crystal, a red-shade blue pigment; and the somewhat finer beta crystal, a green-shade blue pigment (which should not be confused with the halogenated phthalocyanine green pigments). The alpha red-shade phthalocyanine blue is metastable, and unless properly treated will tend to revert to the more stable beta green shade in the presence of strong solvents.

Early phthalocyanine blues, such as Pigment Blue 15, were not treated as were the later alpha red-shade pigments (Pigment Blue 15.1 and 15.2), and would, over time in the presence of strong solvents, exhibit a color drift from a redder to a greener hue. Thus the same paint, depending upon its age (interval between manufacturing and application), may exhibit a greener color than would the same paint applied immediately following its manufacture.

Unaware of the exact color of the aircraft, Mr Hara continues, it is also possible that in pastel light blue tints, the shift to green may relate to resin color. The Japanese were probably using alkyds or phenolic varnishes as binders for coatings on their airplanes in the 1930s and 1940s. The phenolic especially would tend to yellow badly on ultraviolet exposure, and this might lead to a greening of a pastel blue color, particularly in a translucent application (Both varnish and clear lacquer have been mentioned as binders from different sources, leaving uncertainty if this is intended to be one in the same, or separate instances with the same results.).

This author has observed that the color of aotake on actual aircraft appears to be 'greenish" where it has been exposed to the influences of light and air, but "bluish" where exposure has been limited, for example under close fitting components of the aircraft. We soon have the answer. Mr Ford concludes in his report:

> Since the lacquer is a yellowish color, it is likely that the green appearance of the paint is due to the saturation of the blue dye with the yellow lacquer (because blue plus yellow is green), and that the yellow color of the lacquer is due to aging and exposure to light and air.

Therefore, the location of aotake on an aircraft determines the degree of exposure and helps to determine the amount of yellowing of the phenolic lacquer with the passing of time. This could explain why under different lighting conditions aotake appears to take on a different intensity of shade and color.

The last bit of color information comes again from Mr Okano of Mitsubishi. The Manual for Zero Model 11 and 21 indicates "transparent paint (light blue color) is used for internal surface, and special paint for light metal (gray rat color) is used for outer surface which was polished." And furthermore, the book titled "Paint and Marking for Zero" on page 127 mentions "...as indicated in the Manual for Model 11 and 21, inside of body is, in general, painted by light blue transparent paint called 'Aotake.'" Mr Okano continues that some airplanes produced later in WWII did not have such paint. Also, there were Aotake No 1 and No 2, etc, and the color ranged from indigo blue to light green, and even included browns. But in all cases, for what we refer to as aotake, it is transparent.

In the final analysis then, what is aotake? It is a paint generally intended to be used where light fastness and chemical stability are required. But what the Japanese did not realize at the time, or saw of little importance, is that the severe degradation of phenolic varnish under exposure to ultraviolet light changes its color to yellow. It does not matter whether the aotake was exposed or protected, more green or more blue, or even another color, it accomplished its mission; salt-water corrosion protection, and even more, it has kept us guessing for more than 50 years. But readers of this book may have opinions different from those expressed here; if so, the author would like to know.

Index

(add 1 page to each index entry)

Japanese Aircraft by Manufacturer

Japanese Aircraft by Allied Code Name

Japanese Aircraft without Allied Code Name

Non-Japanese Aircraft

Nakajima Ki-84 a/b HAYATE
in Japanese Army Air Force Service
Richard M. Bueschel

Richard Bueschel revises and updates his classic series of books on Japanese Naval and Army Air Force aircraft of World War II. The JAAF Nakajima Ki-84 HAYATE (Frank) is presented in this volume. All variations and markings are covered in this fifth book in a multi-volume series.

ISBN: 0764301497
Size: 8 1/2" x 11"
Illustrations: over 100 b/w photographs
Pages: 64 pages
soft cover
$14.95

Nakajima Ki-49 DONRYU
in Japanese Army Air Force Service
Richard M. Bueschel

The JAAF Nakajima Ki-49 DONRYU (Helen) is presented in this volume. All variations, markings and units are covered in this, the seventh in a multi-volume series.

ISBN: 0764303449
Size: 8 1/2" x 11"
Illustrations: over 90 b/w photographs
Pages: 64 pages
soft cover
$14.95